T0397783

Asymptotics for Fractional Processes

Asymptotics for Fractional Processes

James Davidson

OXFORD
UNIVERSITY PRESS

Great Clarendon Street, Oxford, OX2 6DP,
United Kingdom

Oxford University Press is a department of the University of Oxford.
It furthers the University's objective of excellence in research, scholarship,
and education by publishing worldwide. Oxford is a registered trade mark of
Oxford University Press in the UK and in certain other countries

Published in the United States of America by Oxford University Press
198 Madison Avenue, New York, NY 10016, United States of America

British Library Cataloguing in Publication Data

Data available

Library of Congress Control Number: 2025900063

ISBN 9780198955177

DOI: 10.1093/9780198955207.001.0001

Contents

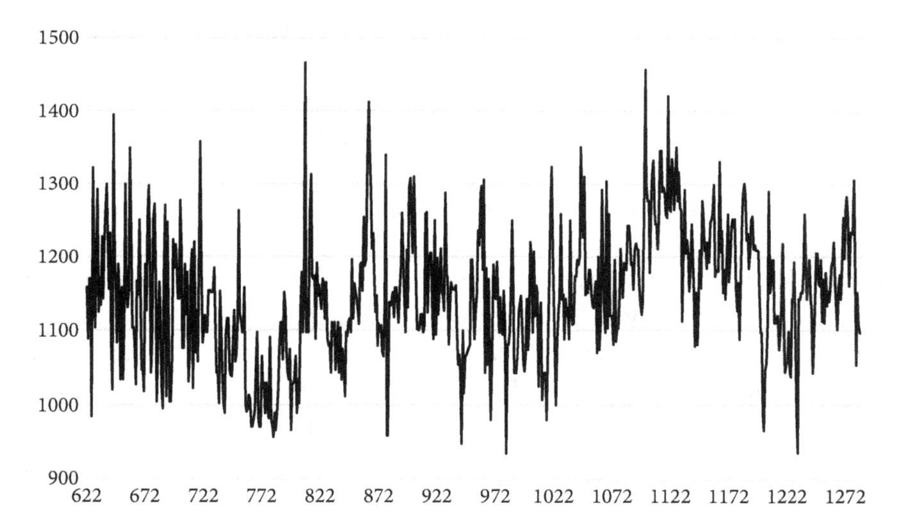

Annual depth minima of the Nile, AD622–1284
Source: Data from [79].

Preface

The object of this book is to develop an approach to the large-sample analysis of so-called fractional partial-sum processes, featuring long memory increments. Long memory in a time series, equivalently called strong dependence, is commonly defined to mean that the autocovariance sequence is nonsummable. These processes depend on a single parameter to measure the degree of persistence, denoted d in the econometrics literature although in statistics the symbol H to denote $d + \frac{1}{2}$ is the well-established usage. Long memory means that d is positive, while negative d is a feature of a special type of short memory known as antipersistence, in which the autocovariance sequence sums to zero. Antipersistent processes are treated here in parallel with the long memory case.

An extensive literature on long memory takes the route of harmonic analysis, the spectral density depending on d in a characteristic manner to provide the basis for various semiparametric estimators. By contrast, the time domain focus adopted in this book is built around and depends on a linear moving average structure. This is the simplest, not to say the most feasible approach, but with Wold's theorem in mind it is also hard to think of a more plausible way to model the decay of autocovariances. It is the fact that the fractional model provides a natural generalization of the unit root autoregression that has captured the imagination of many researchers.

Two major topics are treated. Chapters 1, 2, and 3 show the weak convergence of certain normalized partial sums to fractional Brownian motion, otherwise known as fBm. This is an almost surely continuous Gaussian process that, unlike regular Brownian motion, exhibits correlated increments. The proof of the functional central limit theorem uses well-established methods after reordering the partial sum to group the coordinates of the driving shock process. Then the main difference from the conventional analysis of unit root processes converging to regular Brownian motion is that the increments are heteroscedastic. There are also issues with moment restrictions in the antipersistent case.

Chapters 4, 5, and 6 treat different aspects of the most technically challenging problem, the limiting distribution of stochastic integrals where both the integrand and the integrator processes exhibit either long memory or antipersistence. The approach is different from the conventional analysis of unit root processes since these are not Itô integrals, although the limit distributions feature the sums of distinct Itô-distributed components. Chapter 7 reviews applications of the theory to regression with fractional processes.

Two versions of the results are given, assuming first that the driving shocks are independent and identically distributed and then that they possess a non-parametric form of weak dependence (near-epoch dependence on a mixing process). Dealing with short-range dependence adds a substantial layer of complication and this arrangement allows the reader to decide at any point whether the additional effort of dealing with the general case is worthwhile. Since no restrictions are placed on the models studied other than linearity and the behaviour of the moving average coefficients at long range, even with i.i.d. shocks the theory embraces an extensive class of time series processes. It is really only when nonlinear features such as conditional heteroscedasticity have to be confronted that these results would prove strictly inadequate. In large samples, the replacement of the shock variance by its long-run counterpart is in most cases the only consequence of the generalization. Extensions to cover the dependent shocks case are gathered in Chapter 8.

The final two chapters are included mainly to provide context, with accounts of some closely related results. The chief intention here is to inform readers curious to compare alternative methodologies and alternative generalizations of the unit root class of nonstationary processes. Chapter 9 sets out the essentials of the harmonizable representation of fractional processes, the main revelation being that the asymptotic results obtained in this framework match those already found in the time domain analysis. The same limit properties can be demonstrated in very different ways. Chapter 10 is about local-to-unity autoregression, in which the limit distribution is an Ornstein-Uhlenbeck process instead of fractional Brownian motion.

There are two appendices. Appendix A contains some results in probability theory not directly related to fractional models but needed for deriving the limit distributions. Appendix B lists various trigonometric and related identities arising in the text for the benefit of those desiring an aide mémoire, together with a number of integrals to be encountered whose solutions are not elementary.

These chapters were originally conceived as being a part of the author's *Stochastic Limit Theory* 2nd Edition ([14]), henceforth abbreviated as SLT. A free-standing contribution has proved the better option but SLT plays an important role as a source for certain necessary results, that accordingly can be cited rather than reproved. Readers are referred there for certain theorems arising in Chapters 3 and 6, in particular. Citations of other relevant proofs are supplied in footnotes at various points. It may be useful for readers to know that SLT covers most of the topics that they are assumed in these pages to possess knowledge of.

A note about format and presentation. Numbered assumptions, theorems, and lemmas are terminated by □ unless followed immediately by a proof, while proofs are terminated by ■. References to these items appear in the text in boldface while references to chapter sections have the prefix §. Certain frequently used notations are relatively nonstandard and so should be defined. Thus, in a relation involving sequences x_n and y_n, the notation $x_n \ll y_n$ replaces the more commonly used $x_n \leq Cy_n$ for all n, for a positive constant C. The notation $x_n \simeq y_n$ indicates that there are constants $C_1 > 0$ and $C_2 < \infty$ such that $C_1 \leq x_n/y_n \leq C_2$ for all n, generalizing the standard tilde notation $\sim$ denoting that $x_n/y_n \to 1$. Braces are most commonly used to denote the sequence or array whose generic element is enclosed, as in $\{x_i\}$, but are also used to define sets. The indicator of a set A is written 1_A. $[x]$ is the floor function of x, the largest integer below x. The abbreviation a.s. stands for 'almost sure' or 'almost surely' and equivalently, 'with probability 1'. Finally, '$\to_{\rm d}$' and '$\to_{L_2}$' denote, respectively, convergence in distribution and in mean square.

Some of the results presented here are based on material from published journal articles by myself and coauthors Robert de Jong and Nigar Hashimzade. The contributions of these colleagues to the enterprise are most gratefully acknowledged, as is that of Morten Nielsen, who most helpfully commented on an early draft. I have also received some valuable comments from anonymous referees. Not least, I must thank Xiaoyu Li for her careful reading of the typescript. The remaining errors are most definitely mine.

James Davidson
University of Exeter

Chapter 1
The Fractional Model

There is still no better illustration of the long memory phenomenon than the first example to be identified in the literature. The frontispiece shows the annual minima of depth measurements of the River Nile, taken at the Roda gauge near Cairo, for the 663 years from AD 622 to AD 1284, as recorded in Omar Tousson's history of the Nile ([79]). Famously, these data were studied by the hydrologist Harold Edwin Hurst (1880–1978) who was engaged in the design of the Aswan High Dam. The problem was to specify a dam high enough (but no higher) such that the lake to be created behind it would neither empty nor overflow under variations in rainfall at the river's source, given a constant rate of discharge downstream. This question was easily answered if it could be assumed that the measured flow generated an independent, or at worst weakly dependent, sequence over time. The distribution of the range of a centred and normalized Brownian motion over a fixed period is easily calculated, and would provide a good approximation if the period were long enough. What Hurst showed decisively by means of his well-known rescaled range (R/S) test (see [41], [51], [76]) was that this distribution was the wrong one.

Two features stand out in the plot of what must be thought of as an annual record of precipitation in central Africa. The first is the complete absence of trend over this lengthy historical period. This is, from any reasonable perspective, a stationary time series. But at the same time the persistence of the variations is very striking. The median of the series is 1148 centimetres. In the 86 years 720–806, this level was exceeded only twice. At the other extreme, in the 33 years 1099–1132 the level did not once fall below the median. The Egyptians living through these prolonged periods of drought and flood, respectively, may have thought with some reason that 'the climate is changing'. The full record says otherwise.

It is to build models that can describe and predict series of this type that the theory of fractionally integrated processes has been developed. Early contributions to this literature include the classic papers of Clive Granger ([32], [31]), Jonathan Hosking ([36], [37]), and Peter Robinson ([66]). A useful general account is the monograph by Jan Beran ([4]) and prominent among more

Asymptotics for Fractional Processes. James Davidson, Oxford University Press. © James Davidson (2025).
DOI: 10.1093/9780198955207.003.0001

recent contributions to the statistical analysis are the comprehensive monographs by Giraitis et al. ([28]), Pipiras and Taqqu ([58]), and Beran et al. ([5]), the latter featuring an extensive bibliography. Among other valuable references are the books by Palma ([56]), Hassler ([34]), Samorodnitsky ([70]), and Doukhan et al. ([24]).

1.1 The Model

The model is built around the paradigm of the linear moving average driven by a zero mean innovation sequence $\{u_i\}_{i=-\infty}^{\infty}$, otherwise known as the shock sequence. The basic setup gives the form of the process at date i as

$$x_i = \sum_{j=0}^{\infty} b_j u_{i-j} \tag{1.1}$$

where $b_0 = 1$. The structure defining a fractional (long memory) process is expressed by the equivalence

$$b_j \sim dj^{d-1} L(j) \tag{1.2}$$

as $j \to \infty$, with $d > 0$, where d is the fractional parameter of the process. The sequence $\{L(j)\}_{j=0}^{\infty}$ is positive and either constant or slowly varying at infinity, having the property $L(ax)/L(x) \to 1$ as $x \to \infty$ for any $a > 0$. Powers of $\log(x)$ are the familiar examples of slow variation. Only the behaviour of the sequence at long range is specified by (1.2), b_j being allowed to take arbitrary form when j is finite. The distribution of the process $\{x_i, -\infty < i < \infty\}$ is well defined only if d is not too large, as detailed in §1.2.

Subject to these considerations, nothing needs to be added to (1.1)+(1.2) to define what is often called a fractionally integrated process, although strictly this terminology is reserved for the special case to be examined in §1.3. Maximum likelihood estimation of a parametric model conforming to this setup, applied to the Nile minima series, gives a value of d in the vicinity of 0.4 (see [19]). The role of the multiplier d in (1.2) will become clear later on.

With $d > 0$ the coefficient sequence $\{b_j\}_{j=0}^{\infty}$ is nonsummable. This is the property that characterizes models of the form (1.1) as long memory and equivalently, as strongly dependent. Because of nonsummability the assumption $\mathrm{E}(u_i) = 0$ is essential, since with a nonzero mean included the sum in (1.1) would diverge. A fractional process with nonzero mean μ can exist, but it must take the form $x_i + \mu$ where x_i follows (1.1)+(1.2).

Another leading implication of strong dependence is the failure of the conventional central limit theorem. Thus,

$$\frac{1}{\sqrt{n}}\sum_{i=1}^{n} x_i = \sum_{j=0}^{\infty} b_j \left(\frac{1}{\sqrt{n}} \sum_{i=1}^{n} u_{i-j} \right) = \sum_{j=0}^{\infty} b_j Z_n(j) \tag{1.3}$$

(say) where under various conditions on the shock sequence $\{u_i\}$, such as serial independence, the variables $Z_n(j)$ would approach the same Gaussian limit for each j, as $n \to \infty$. In the case $|\sum_{j=0}^{\infty} b_j| < \infty$, this Gaussian property would extend to $n^{-1/2}\sum_{i=1}^{n} x_i$ itself, the only effect being a change of scale. However, under long memory, $|\sum_{j=0}^{\infty} b_j| = \infty$ and the variance of the sum diverges. The 'square root rule' of normalization by $n^{-1/2}$ does not apply here.

The permitted inclusion of slowly varying components in (1.2) involves something of a sleight of hand, because to do the asymptotics requires that the L function be included in the normalizing divisor to obtain the limit distribution of a statistic, and so made to disappear. If neglected, a slowly varying drift would render the limit distributions invalid although hopefully by a small enough amount that finite-n approximations would be useful. Such components are rarely if ever specified in empirical applications and there is a sense in which they are mainly window-dressing. The mathematics would be simplified by their omission, but if slowly varying components do exist it at least tells the practitioner what to do about them.

There is the larger issue that the parameter d must also be known to make use of the limit results. Estimation falls outside the subject matter covered in this book but a range of procedures are available to estimate d from the data. Among innumerable references see for example [27], [54], [67], [74], [35], and [19].

1.2 Shocks and Dependence

Since no restriction beyond (1.2) is imposed on the moving average coefficients, allowing them to take arbitrary forms for finite lags, assuming that the shocks $\{u_i\}$ are independently distributed does no more than confine attention to linear forms of serial dependence, less restrictive than might be supposed (see §1.5). In calculations involving second moments, much simplicity in exposition is thereby achieved. However, for a complete picture the effects of dependent disturbances do need to be known. Therefore, two alternative sets of assumptions are invoked at different points in the development.

1.1 Assumption The sequence $\{u_i, -\infty < i < \infty\}$ is identically and independently distributed (i.i.d.) and L_r-bounded for $r \geq 2$, with $\mathrm{E}(u_i) = 0$ and $\mathrm{E}(u_i^2) = \sigma_u^2$ where $0 < \sigma_u^2 < \infty$. □

1.2 Assumption The sequence $\{u_i, -\infty < i < \infty\}$ is

(a) strictly stationary and L_r-bounded for $r \geq 2$, with $\mathrm{E}(u_i) = 0$, $\mathrm{E}(u_i^2) = \sigma_u^2$ where $0 < \sigma_u^2 < \infty$ and for $k > 0$, $\mathrm{E}(u_i u_{i+k}) = \gamma_u(k) = O(k^{-1-\delta})$ for $\delta > 0$.

(b) L_2-near epoch dependent (NED) of size $-\frac{1}{2}$ on either a α-mixing sequence of size $-r/(r-2)$ with $r > 2$, or a ϕ-mixing sequence of size $-r/(2r-2)$. □

Assumption **1.2**(a) and in particular the requirement $\delta > 0$ specifies that the autocovariances are absolutely summable. This condition defines the boundary between weak and strong dependence, implying that the standard deviations of partial sums of the process grow like $\sqrt{n}$ and hence that $\omega_u^2 < \infty$ where

$$\omega_u^2 = \lim_{n\to\infty} \frac{1}{n}\mathrm{E}\left(\sum_{j=1}^{n} u_i\right)^2 = \sigma_u^2 + 2\sum_{k=1}^{\infty} \gamma_u(k). \tag{1.4}$$

Assumption **1.2**(b) implies the autocovariance summability condition of **1.2**(a), but this is nonetheless stated explicitly to define notation and focus attention on the important implications of weak dependence.

Since Assumption **1.1** is most convenient for exposition, the strategy adopted here is to collect in one place the generalizations needed to move to Assumption **1.2**. What will be shown in §2.6, §3.5, and Chapter 8 in particular is that although the methods of analysis are different, relaxing the i.i.d. assumption typically involves no change in asymptotic results beyond replacement of σ_u^2 by ω_u^2 in formulae.

The next consideration is the magnitude of d. If $d < \frac{1}{2}$ the sequence $\{b_j\}_{j=0}^{\infty}$ in (1.2) is square-summable, which is a sufficient condition for covariance stationarity of the process. The variance of $\{x_i\}$ with $d < \frac{1}{2}$ under Assumption **1.1** is

$$\sigma_x^2 = \mathrm{E}(x_i^2) = \sigma_u^2 \sum_{j=0}^{\infty} b_j^2 \backsimeq \sum_{j=0}^{\infty} j^{2d-2} L(j)^2 < \infty. \tag{1.5}$$

The cited estimate of d for the Nile minima series falls in the covariance stationary region although not too far from its boundary, which is what the eyeball analysis of the series would also suggest. The autocovariances $\gamma_k = \mathrm{E}(x_i x_{i+k})$ exist similarly[1] and

$$\gamma_k = \sigma_u^2 \sum_{j=0}^{\infty} b_j b_{j+k} \backsimeq \sum_{j=k+1}^{\infty} j^{2d-2}\left(1 - \frac{k}{j}\right)^{d-1} L(j)^2 = O(k^{2d-1}). \qquad (1.6)$$

Weak dependence is strictly defined by the property of summable autocovariances and it is apparent from (1.6) that under stationary strong dependence the autocovariances are nonsummable.

Even when stationary these processes fail to satisfy useful mixing and near-epoch dependence (NED) conditions.[2] Under Assumption **1.1**, the L_2-NED criterion for x_i can be evaluated as

$$\|x_i - \mathrm{E}_{i-m}^{i+m} x_i\|_2 = \sigma_u \left(\sum_{j=m+1}^{\infty} b_j^2 \right)^{1/2} = O(m^{d-1/2}).$$

This does vanish with $d < \frac{1}{2}$, but not fast enough for the central limit theorem to operate. With u_i dependent under Assumption **1.2**, no NED property of any kind can be demonstrated for x_i. The usual nonparametric techniques for modelling dependence are not available here.

In the case $d \geq \frac{1}{2}$, the process exhibits too much persistence to possess a variance and is covariance-nonstationary. However, if $d < 1$ then $b_j \to 0$ as $j \to \infty$ and although it has no moments the distribution of $\{x_i\}$ is stationary in the sense of independence of initial conditions. Its behaviour is therefore distinct from the case $d = 1$ which under Assumption **1.1** defines a random walk, a nonstationary distribution that is well defined only when it has a finitely remote start date. This matches the unit root case of the ARMA model class so that fractionals provide an alternative class within which the 'unit root' model is embedded, although take care to note that d is not the root of any equation.

An integrated process, of which the random walk is the most familiar case, is one formed by cumulation of stationary increments from some initial date.

[1] This object arises frequently, so for economy of notation γ_k stands here for $\gamma_x(k)$.

[2] Consider the sufficient mixing condition under linearity of SLT Theorem 15.9. Under (1.2), there is no positive value of d to satisfy condition (b) of that theorem.

If x_i is stationary long memory with parameter $d \in (0, \frac{1}{2})$, then $S_i = \sum_{j=1}^{i} x_j$ is a nonstationary fractional process having fractional parameter equal to $1+d \in (1, \frac{3}{2})$. The distribution of an integrated process depends on the distribution of the underlying stationary process and the cumulation start date, which can be set at $i=1$ without loss of generality. The cumulation procedure can be iterated so that in principle d can take any positive value.

In common with the ARMA class, the fractional model is built around the linear moving average structure of (1.1). The one model feature intrinsically associated with long memory is the parameter d and according to the equality in (1.6), the autocovariance sequence $\{\gamma_k\}_{k=0}^{\infty}$ appears to provide a complete specification of the memory properties of x_i in (1.1). It is reasonable to ask whether the linear framework might be dispensed with and the long memory property specified directly by the autocovariances, or more conveniently in terms of the spectral density (the Fourier transform of the autocovariances). Some details of this approach are given in Chapter 9. Working with the spectrum, dependence on d can be expressed semiparametrically.

Whether there is a gain in generality to be achieved is best answered by considering the Wold decomposition theorem, according to which a stationary and nondeterministic time series always has a linear moving average representation with *white noise* shocks; that is, covariance stationary and uncorrelated with zero mean. Except in the Gaussian case, this is not the same thing as i.i.d. as specified in Assumption **1.1**. However, shock dependence that does not contradict the white noise assumption entails uncorrelatedness and hence must be nonlinear, prominently including conditional heteroscedasticity of the ARCH and GARCH type,[3] which is permitted under Assumption **1.2** (see [15]). Defining long memory as a property of the second moments of the joint distribution of a process is virtually equivalent to specifying (1.1)+(1.2) under Assumption **1.1**. It is also nearly equivalent under Assumption **1.2**, in the sense that the cases ruled out of the comparison must feature nonlinear dependence.

A moving average framework is also adopted in generalizations of GARCH models of volatility that are sometimes called fractional, although more properly should be described as featuring hyperbolic memory. In these models, while the process itself may be white noise the centred squares are modelled as a fractional linear process. However, as shown in [16] for example the dynamic properties of these models differ in subtle and important ways from fractional models of the level.

[3] See for example SLT §18.6.

1.3 Fractional Integration

The best-known mechanism for inducing the behaviour summarized in (1.2) is represented by the recursion

$$b_j = \frac{j+d-1}{j} b_{j-1}, j > 0 \tag{1.7}$$

where $b_0 = 1$. (No slowly varying component in this instance.) These b_j are the coefficients of the binomial expansion of the so-called fractional integral

$$(1-B)^{-d} = \sum_{j=0}^{\infty} b_j B^j \tag{1.8}$$

where B denotes the backshift operator (or lag operator) defined by the relation $Bu_i = u_{i-1}$, so that

$$x_i = (1-B)^{-d} u_i \tag{1.9}$$

is a case of (1.1). The solution of the fractional integral as an infinite-order lag polynomial follows from Newton's generalized binomial theorem, which says that for $|t| < 1$,

$$(1+t)^r = \sum_{j=0}^{\infty} \frac{(r)_j}{j!} t^j \tag{1.10}$$

where r is any real number and

$$(r)_j = r(r-1)\cdots(r-j+1). \tag{1.11}$$

The latter function is known as a falling factorial and $(\cdot)_j$ is called the Pochhammer symbol.

The relation $j! = \Gamma(j+1)$ where Γ denotes the gamma function (see (B.12) of Appendix B) follows by recursive application of (B.13) starting from $\Gamma(1) = 1$. While $(r)_j = r!/(r-j)!$ is only true when r is a positive integer and $r \geq j$, (B.13) shows that the equality

$$(r)_j = \frac{\Gamma(r+1)}{\Gamma(r-j+1)} \tag{1.12}$$

holds for real r. The recursion in (1.7) defines the series

$$b_j = \frac{(d+j-1)_j}{j!} = \frac{\Gamma(j+d)}{\Gamma(d)\Gamma(j+1)} = \frac{d\Gamma(j+d)}{\Gamma(d+1)\Gamma(j+1)} \tag{1.13}$$

for $j \geq 1$. In view of the fact that

$$(-r)_j = (-1)^j (r + j - 1)_j$$

it can be verified from the first equality of (1.13) that

$$\frac{(-d)_j}{j!} = b_j (-1)^j.$$

Setting $r = -d$ and replacing t by $-B$, the expansion in (1.10) assumes the form of (1.8). It is clear why with $d > 0$ the process generated by (1.8) is referred to as 'fractionally integrated', since the 'fully integrated' case with $d = 1$ corresponds to the autoregressive unit root. It follows from (1.11) and (1.12) for fixed j and increasing r that for a fixed value of a,

$$\frac{\Gamma(r) r^a}{\Gamma(r + a)} \to 1 \text{ as } r \to \infty. \tag{1.14}$$

Therefore (1.13) implies

$$b_j \sim \frac{d j^{d-1}}{\Gamma(d + 1)} \tag{1.15}$$

which may be compared with (1.2). The class of models having lag structure (1.8) with $d \in (0, \frac{3}{2})$ provide a continuum of dependence properties within which the unit root case is embedded.

A useful feature of special case (1.13) is that the autocovariances can be evaluated exactly. The following derivation adapts Lemma 1 of [38].

1.3 Theorem

$$\gamma_k = \sigma_u^2 \frac{\Gamma(1 - 2d)\Gamma(d + k)}{\Gamma(1 - d + k)} \frac{\sin \pi d}{\pi}. \tag{1.16}$$

Proof Substituting from (1.13) into the formula in (1.6) gives

$$\begin{aligned} \gamma_k &= \sigma_u^2 \sum_{j=0}^{\infty} \frac{\Gamma(j + d)}{\Gamma(d)\Gamma(j + 1)} \frac{\Gamma(j + k + d)}{\Gamma(d)\Gamma(j + k + 1)} \\ &= \sigma_u^2 \frac{\Gamma(d + k)}{\Gamma(d)\Gamma(k + 1)} F(d, k + d; k + 1; 1) \\ &= \sigma_u^2 \frac{\Gamma(1 - 2d)}{\Gamma(d)\Gamma(1 - d)} \frac{\Gamma(d + k)}{\Gamma(1 - d + k)} \end{aligned} \tag{1.17}$$

where the second equality of (1.17) notes the match with the hypergeometric series defined in (B.22) and the third is got by simplification after applying identity (B.23). The formula in (1.16) is lastly obtained by use of (B.15). ∎

Another version of formula (1.16) sometimes cited is

$$\gamma_k = \sigma_u^2 \frac{\Gamma(1-2d)(-1)^k}{\Gamma(1-d+k)\Gamma(1-d-k)}.$$

Switching between the two is a simple application of identities (B.15) and (B.5).

Applying (1.14) shows that $\gamma_k = O(k^{2d-1})$, agreeing with the calculation in (1.6). The case of particular interest, easily shown using (B.15) after setting $k = 0$, is

$$\gamma_0 = \sigma_x^2 = \sigma_u^2 \frac{\Gamma(1-2d)}{\Gamma(1-d)^2}. \tag{1.18}$$

As expected, this formula diverges as d approaches the stationarity boundary of $\frac{1}{2}$.

1.4 Antipersistence

Antipersistent processes are certain cases of (1.1)+(1.2) having fractional parameter $d < 0$. This latter condition is not by itself of special interest since it merely implies summable lag coefficients and hence, weak dependence in general. The distinguishing feature of the antipersistent process is that the lag coefficients sum to zero.

A process with this property is created by forming the differences of a process whose moving average coefficients converge to zero. In particular, an antipersistent process having $d \in (-\frac{1}{2}, 0)$ is formed as the differences of a covariance-nonstationary process having fractional parameter equal to $d + 1$, in the range $(\frac{1}{2}, 1)$. To show this, suppose that $y_i = \sum_{j=0}^{\infty} a_j u_{i-j}$ and let $x_i = y_i - y_{i-1} = \sum_{j=0}^{\infty} b_j u_{i-j}$. Then, $b_0 = a_0 = 1$ and $b_j = a_j - a_{j-1}$ for $j > 0$. If $a_j \to 0$ as $j \to \infty$ then, evidently,

$$1 + (a_1 - 1) + (a_2 - a_1) + (a_3 - a_2) + \cdots = 0.$$

If in particular $a_j \sim j^{(d+1)-1}L(j)$ with $d < 0$, then as $m \to \infty$,

$$\sum_{j=0}^{m} b_j = a_m \sim m^d L(m). \tag{1.19}$$

Also, as $j \to \infty$,

$$j^d L(j) - (j-1)^d L(j-1) \sim d j^{d-1} L(j) \tag{1.20}$$

matching the representation in (1.2). In the case of the fractional integral defined by (1.13), the relation in (1.20) holds identically since applying (B.13) gives

$$\begin{aligned}\frac{\Gamma(j+(d+1))}{\Gamma(d+1)\Gamma(j+1)} - \frac{\Gamma((j-1)+(d+1))}{\Gamma(d+1)\Gamma((j-1)+1)} &= \frac{\Gamma(j+d)}{\Gamma(d+1)\Gamma(j)}\left(\frac{j+d}{j} - 1\right) \\ &= \frac{\Gamma(j+d)}{\Gamma(d)\Gamma(j+1)}.\end{aligned}$$

According to the first equality in (1.6), under Assumption **1.1** (noting $\gamma_k = \gamma_{-k}$ under stationarity) another implication is that

$$\sum_{k=-\infty}^{\infty} \gamma_k = \sigma_u^2 \left(\sum_{j=0}^{\infty} b_j\right)^2 = 0. \tag{1.21}$$

Antipersistent processes must exhibit negative autocorrelation. In (1.16), $\gamma_k < 0$ for every $k > 0$ if $d \in (-\frac{1}{2}, 0)$. The fact that the γ_k sum to zero over the range $k = -\infty, \dots, \infty$ is less obvious from the formula, but this must follow from (1.21) and (1.19).

Like cumulation, the differencing operation can be iterated so that in principle d can assume any negative value, although differencing a process loses the initial value and so cannot be undone by cumulation in the same way that cumulation can be undone by differencing. A process formed as the differences of a covariance stationary process and so having a fractional parameter smaller than $-\frac{1}{2}$ is said to be overdifferenced, because the operation takes the data series outside of the class of processes amenable to asymptotic analysis. The boundaries $d < \frac{1}{2}$ and $d > -\frac{1}{2}$ together delineate the set of processes whose normalized partial sums converge to a.s. continuous Gaussian limits under suitable conditions on the shocks, as is to be shown in the next two chapters. The restriction $|d| < \frac{1}{2}$ is set for all the results to be derived in the sequel.

A related concept needing to be carefully distinguished from overdifferencing is noninvertibility. An invertible process having the linear moving average form of (1.1) is one that allows the innovation process $\{u_i\}$ to be solved from the observables $\{x_i\}$. A convenient model for understanding these issues is the fractional integral from (1.9), since inverting the polynomial is then straightforward. The autoregressive representation of the process is $(1 - B)^d x_i = u_i$, or equivalently

$$x_i = \sum_{j=1}^{\infty} \phi_j x_{i-j} + u_i \tag{1.22}$$

whereas in (1.15) the lag coefficients are

$$\phi_j = -\frac{\Gamma(j-d)}{\Gamma(-d)\Gamma(j+1)} = O(j^{-d-1}).$$

If $d \leq -1$, this coefficient sequence is either non-vanishing or divergent as $j \to \infty$, which implies that u_i cannot be solved from (1.22) and the process $\{x_i\}_{i=-\infty}^{\infty}$ does not have a well-defined distribution.

For this reason, the moving average counterpart of this case in (1.9) with $d \leq -1$ is called noninvertible. Another noninvertible case is an ARMA process with moving average root of unity. Fractional processes with $d \in (-1, -\frac{1}{2}]$ are invertible in spite of being overdifferenced, since $\phi_j \to 0$ as $j \to \infty$ and the representation in (1.22) is well defined. For reasons that will become clear in the development, they are nonetheless excluded from consideration in the asymptotic analysis of partial sums.[4]

1.5 Linearity

The pure fractional model (1.9) depending on a single parameter is somewhat restrictive, but if the route to a parametric generalization is to let the shocks be a weakly dependent process, staying within the linear paradigm of (1.1)+(1.2) is the natural consideration. Consider replacing u_i in (1.9) by v_i, itself having the moving average representation $v_i = \varphi(B)u_i$ where $\{u_i\}$ satisfies Assumption **1.1** and $\varphi(B) = \sum_{j=0}^{\infty} \varphi_j B^j$ denotes a polynomial in the lag operator with absolutely summable coefficients. A natural choice of model is the ARMA(p, q), for which $\varphi(B) = \theta(B)/\phi(B)$ where $\phi(B)$ and $\theta(B)$ are respectively stable and invertible lag polynomials of finite orders p and q. The rate of decay of the lag coefficients is then geometric. The so-called ARFIMA(p, d, q) (autoregressive fractionally integrated moving average) model with representation

[4] See [55] and [8] for alternative analyses of this question.

$$\phi(B)(1-B)^d x_i = \theta(B)u_i \tag{1.23}$$

is a popular choice in empirical studies of long memory.

The following result shows that the only effect of this extension on the long-range behaviour of the process is a change of scale. This generalizes beyond the fractional integral in (1.9) by just assuming (1.2). The slowly varying component is omitted here for the sake of simplicity but its inclusion would change nothing material.

1.4 Theorem If $a(B) = b(B)\varphi(B)$ denotes a composite lag polynomial, where $b_j \sim Kj^{d-1}$ for $|d| < \frac{1}{2}$ and a constant K and $|\varphi_j| = O(j^{-1-\delta})$ for $\delta > 0$, then

$$a_j \sim \varphi(1)Kj^{d-1} \tag{1.24}$$

as $j \to \infty$.

Proof Gathering terms of the product with matching powers of B shows that

$$a_j = \sum_{k=0}^{j} \varphi_k b_{j-k}$$

for each $j \geq 0$. Choose η from the interval $(1/(1+\delta), 1)$ and so write

$$a_j \sim K\sum_{k=0}^{j-1} \varphi_k (j-k)^{d-1} = Kj^{d-1}\left(\frac{j-j^\eta}{j}\right)^{d-1} \sum_{k=0}^{j-1} \varphi_k \left(\frac{j-k}{j-j^\eta}\right)^{d-1}. \tag{1.25}$$

Break up the sum on the right-hand side of (1.25) as

$$\left(\sum_{k=0}^{[j^\eta]-1} + \sum_{k=[j^\eta]}^{j-1}\right)\varphi_k \left(\frac{j-k}{j-j^\eta}\right)^{d-1} = A(j) + B(j). \tag{1.26}$$

Since $(j-k)/(j-j^\eta) \to 1$ as $j \to \infty$ for any fixed $k < [j^\eta]$, it is immediate that $A(j) \to \varphi(1)$. In view of the fact that $(j-j^\eta)/j \to 1$, the proof is completed by showing $|B(j)| \to 0$.

After substituting $|\varphi_k| \ll k^{-1-\delta}$, the required argument is to be found in a different context in Lemma **8.2** on page 146. Specifically, putting $m = j - k$,

$$
\begin{aligned}
|B(j)| &= \sum_{m=1}^{j-[j^\eta]} \left(\frac{m}{j-j^\eta}\right)^{d-1} |j-m|^{-1-\delta} \\
&\leq j^{-\eta(1+\delta)}(j-j^\eta)^{1-d} \sum_{m=1}^{j-[j^\eta]} m^{d-1} \\
&\ll j^{1-\eta(1+\delta)}
\end{aligned}
$$

where the first inequality uses $j - m \geq j^\eta$ and the second is by integral approximation. $B(j)$ vanishes as $j \to \infty$ since $\eta(1+\delta) > 1$. ∎

Chapter 2
Fractional Asymptotics

From the point of view of inference the main interest in fractional processes is with the associated partial sum processes, since it is the time aggregation that gives rise to known distributions in the limit. The familiar case of Brownian motion B has the feature that $B(at) \sim_{\mathrm{d}} a^{1/2}B(t)$ for $t \geq 0$ and $a \geq 0$, where $\sim_{\mathrm{d}}$ denotes equivalence in distribution. This is an example of the self-similarity property of a process, by which the distribution is preserved under changes of scale with suitable normalization. John Lamperti ([43]) first suggested extending the class of self-similar processes X to cases of the type $X(at) \sim_{\mathrm{d}} a^{\alpha}X(t)$, with the self-similarity index α falling either above or below $\frac{1}{2}$. The same idea was explored by Benoit Mandelbrot ([47], [48]) who defined fractional Brownian motion, with the abbreviation fBm, to be the Gaussian case of this generalization. For an excellent survey of its properties by Murad Taqqu, see [78]. As previously remarked, the symbol H (after Hurst) is commonly used for the self-similarity index in the statistical literature, corresponding to $d + \frac{1}{2}$, so that $0 < H < 1$ is the range of interest matching $|d| < \frac{1}{2}$.

2.1 Fractional Brownian Motion

Fractional Brownian motion is the a.s. continuous stochastic process $X\colon [0,1] \mapsto \mathbb{R}$ that is represented in the time domain by

$$X(t) = \int_0^t (t-\xi)^d \mathrm{d}U(\xi) + \int_{-\infty}^0 \left((t-\xi)^d - (-\xi)^d\right) \mathrm{d}U(\xi),\ t \in [0,1] \qquad (2.1)$$

where $|d| < \frac{1}{2}$ and U denotes a Brownian motion process with variance $\mathrm{E}(U(1)^2) = \sigma_u^2$, where without loss of generality it can be stipulated that $U(0) = 0$. The slightly counter-intuitive notion of $U(\xi)$ evolving from a starting point at $\xi = -\infty$ can be better appreciated by noting that a Brownian motion has independent increments and is accordingly reversible. The distribution on the negative half-line can be paired with that on the positive

Asymptotics for Fractional Processes. James Davidson, Oxford University Press. © James Davidson (2025).
DOI: 10.1093/9780198955207.003.0002

half-line initialized at zero. While $U(t)$ diverges with probability 1 as $t \to \infty$, only the stationary increments $\mathrm{d}U$ enter the formula in (2.1), having square-integrable weights in the second term.

Mandelbrot and van Ness in [48] included the scale factor $1/\Gamma(d+1)$ in their version of formula (2.1) and this would be appropriate if the formulation $b_j \sim dj^{d-1}L(j)/\Gamma(d+1)$ had been given in place of (1.2). The particular case of (1.13) does include this factor, as can be seen in (1.15), but this is the fractional integral formalization of long memory matching the solution of (1.8). In other cases the formulation is optional. See §9.3 for a fuller explanation of these alternatives in the context of the harmonizable representation of the model.

The striking feature of (2.1) is the dependence on the entire past history of the driving process U. Thinking of fBm as the limit process of normalized partial sums, as is to be proved in Chapter 3, weakly dependent increment processes including the ARMA class may take the form of (1.1) and depend on the infinite past, but after normalization the presample contributions to the variation become negligible in the limit. Setting $d = 0$ in (2.1) gives regular Brownian motion $U(t) = \int_0^t \mathrm{d}U(\xi)$, with the second term vanishing. It is only in fractional processes having $d \neq 0$ that the presample contributions persist into the limit distribution. This is also true when $d < 0$, corresponding to the antipersistence case, even though like the ARMA class these are weakly dependent processes with summable autocovariances. This puzzle is resolved by noting the different normalizations required to give rise to the a.s. continuous Gaussian limit process in (2.1).

Sometimes the expression in (2.1) is referred to as 'fractional Brownian motion of Type I' since an alternative formulation known as the Type II case (see [50]) is

$$X(t) = \int_0^t (t-\xi)^d \mathrm{d}U(\xi). \tag{2.2}$$

This is the asymptotic counterpart of the partial sum process driven by the truncated version of (1.1) in which $u_i = 0$ for $i \leq 0$. Since it features non-stationary increments, where the distributions are linked to the time elapsed since what is usually regarded as the first observation of a sample (i.e., unconnected with the phenomenon being modelled, in general) the Type II process is evidently a less satisfactory platform for empirical time series modelling than (2.1) (see [19]). Nonetheless, there is a substantial literature based around these truncated processes so it may be useful to point out that weak convergence to (2.2) can be analysed by a more compact version of the arguments to be deployed here.

2.2 The Variance

An increment of the fBm process of Type I for $0 \leq t < s \leq 1$ has the form

$$X(s) - X(t) = \int_t^s (s-\xi)^d \mathrm{d}U(\xi) + \int_{-\infty}^t \left((s-\xi)^d - (t-\xi)^d\right) \mathrm{d}U(\xi). \quad (2.3)$$

The signature property of fBm is the variance function of these increments.

2.1 Theorem If $|d| < \frac{1}{2}$, $\mathrm{E}(X(s) - X(t))^2 = \sigma_u^2 \Upsilon_d |s-t|^{2d+1} < \infty$ for any $s, t \in [0, 1]$ where

$$\Upsilon_d = \frac{1}{2d+1} + \int_0^\infty \left((1+\tau)^d - \tau^d\right)^2 \mathrm{d}\tau. \quad (2.4)$$

Proof Assume $s > t$ without loss of generality. The orthogonality of Brownian motion increments implies that

$$\mathrm{E}(X(s) - X(t))^2 = \sigma_u^2 \int_t^s (s-\xi)^{2d} \mathrm{d}\xi + \sigma_u^2 \int_{-\infty}^t \left((s-\xi)^d - (t-\xi)^d\right)^2 \mathrm{d}\xi. \quad (2.5)$$

Apply changes of variable $\tau = (s-\xi)/(s-t)$ in the first integral in (2.5) and $\tau = (t-\xi)/(s-t)$ in the second integral, noting $(s-\xi)/(s-t) = \tau + 1$ in that case. The integral in (2.4) exists for $|d| < \frac{1}{2}$. ∎

The increments of fBm are thus shown to be stationary, the variance depending on the width of the interval $|s-t|$, but not on t. As can be verified from the variance formula, self-similarity implies that the distribution of the increment $X(t+\delta) - X(t)$ matches that of $a^{-d-1/2}(X(t+a\delta) - X(t))$ for any $a > 0$.

For Theorem **2.1** to be practically useful, a closed form for the expression Υ_d in (2.4) is desirable. The formula in question is as follows.

2.2 Theorem If $|d| < \frac{1}{2}$,

$$\Upsilon_d = \frac{\Gamma(1-2d)\Gamma(1+d)}{(2d+1)\Gamma(1-d)} \quad \square \quad (2.6)$$

Note the salient fact that $\Upsilon_d = \infty$ in both of the boundary cases, $d = \frac{1}{2}$ and $d = -\frac{1}{2}$. As discussed in greater detail in §2.4, these divergences represent the breakdown of a.s. boundedness of X on the one hand and the failure of a.s. continuity of X on the other.

The proof of Theorem **2.2** utilizes the following reworking of Lemma 5.1 of [18].

2.3 Lemma If $|d| < \frac{1}{2}$,

$$\int_0^\infty \left((\tau + 1)^d - \tau^d\right)^2 \mathrm{d}\tau = \frac{1}{2d+1}\left(\frac{\Gamma(1-2d)\Gamma(1+d)}{\Gamma(1-d)} - 1\right) \tag{2.7}$$

Proof Consider the function of a complex variable

$$\phi(z) = \left((z+1)^d - z^d\right)^2, \; z \in \mathbb{C}.$$

Recalling $\mathrm{e}^{\mathrm{i}\pi} = -1$ define, for a real argument x,

$$z(x) = \begin{cases} x, & x \geq 0 \\ \mathrm{e}^{\mathrm{i}\pi}|x|, & x < 0. \end{cases} \tag{2.8}$$

Then,

$$f(x) = \phi \circ z = \left((z(x) + 1)^d - z(x)^d\right)^2$$

is a well-defined function of a real variable $x \in \mathbb{R}$ whose integral over $[0, \infty)$ is the solution sought.

In the case $0 < d < \frac{1}{2}$, it can be verified both that $f(x) = O(|x|^{2d-2})$ as $x \to \pm\infty$ and that $\lim_{x\to 0} |f(x)| = 1$, so f is integrable for both positive and negative x. In the case $-\frac{1}{2} < d < 0$, f has a singularity at $x = 0$, but $\lim_{x\to 0} \left|f(x)/z(x)^{2d}\right| = 1$ so it is integrable on the nonnegative half-line. It also has a singularity at $x = -1$ but $\lim_{x\to -1} \left|f(x)/(z(x)+1)^{2d}\right| = 1$ so it is likewise integrable on the negative half-line. Therefore, the integral

$$\mathcal{L}_d = \int_{-\infty}^\infty f(x)\,\mathrm{d}x$$

exists for $|d| < \frac{1}{2}$.

Introduce the change of variable $y = -1-x$, with inverse relation $x = -1-y$. For each of the three cases $x > 0$, $-1 < x < 0$, and $x < -1$, it can be verified that $z(x) = \mathrm{e}^{\mathrm{i}\pi}(1 + z(y))$ and also that $1 + z(x) = \mathrm{e}^{\mathrm{i}\pi} z(y)$. Hence,

$$\begin{aligned}\mathcal{L}_d &= \int_{-\infty}^\infty \left((z(x)+1)^d - z(x)^d\right)^2 \mathrm{d}x \\ &= \int_{-\infty}^\infty \left((\mathrm{e}^{\mathrm{i}\pi} z(y))^d - (\mathrm{e}^{\mathrm{i}\pi}(z(y)+1))^d\right)^2 \mathrm{d}y = \mathrm{e}^{2\mathrm{i}\pi d}\,\mathcal{L}_d. \end{aligned} \tag{2.9}$$

It follows from (2.9) that $\mathcal{L}_d = 0$ for $d \neq 0$. Direct inspection shows that $\mathcal{L}_0 = 0$ also holds.

Let $\mathcal{L}_{1d}$, $\mathcal{L}_{2d}$, and $\mathcal{L}_{3d}$ denote the integrals of $f(x)$ over the intervals $(-\infty, -1)$, $(-1, 0)$, and $(0, \infty)$ respectively, so that $\mathcal{L}_{3d}$ is the formula sought and $\mathcal{L}_d = \mathcal{L}_{1d} + \mathcal{L}_{2d} + \mathcal{L}_{3d} = 0$ for all d. Analogously to (2.9) the change of variable $y = -1 - x$ gives

$$\begin{aligned}\mathcal{L}_{1d} &= \int_{-\infty}^{-1} \left((z(x)+1)^d - z(x)^d\right)^2 \mathrm{d}x \\ &= \int_0^{\infty} \left((\mathrm{e}^{\mathrm{i}\pi} z(y))^d - (\mathrm{e}^{\mathrm{i}\pi}(z(y)+1))^d\right)^2 \mathrm{d}y = \mathrm{e}^{2\mathrm{i}\pi d}\mathcal{L}_{3d}.\end{aligned}$$

Next, to find $\mathcal{L}_{2d}$, multiply out $f(x)$ and use the facts that with x in the interval $(-1, 0)$ it is possible to write both $z(x) + 1 = 1 - |x|$ and $z(x) = \mathrm{e}^{\mathrm{i}\pi}|x|$. Thus,

$$\begin{aligned}\mathcal{L}_{2d} &= \int_{-1}^{0} (z(x)+1)^{2d}\mathrm{d}x + \int_{-1}^{0} z(x)^{2d}\mathrm{d}x - 2\int_{-1}^{0} (z(x)+1)^d z(x)^d \mathrm{d}x \\ &= \frac{1-0}{2d+1} + \frac{0 - \mathrm{e}^{\mathrm{i}\pi(2d+1)}}{2d+1} - 2\mathrm{e}^{\mathrm{i}\pi d}\int_0^1 x^d(1-x)^d\,\mathrm{d}x \\ &= \frac{1+\mathrm{e}^{2\mathrm{i}\pi d}}{2d+1} - 2\mathrm{e}^{\mathrm{i}\pi d}B(d+1, d+1)\end{aligned} \tag{2.10}$$

where the last equality notes the definition of the Beta function from (B.14) of Appendix B. Adding over the three intervals yields the relation

$$\mathrm{e}^{2\mathrm{i}\pi d}\mathcal{L}_{3d} + \frac{1+\mathrm{e}^{2\mathrm{i}\pi d}}{2d+1} - 2\mathrm{e}^{\mathrm{i}\pi d}B(d+1, d+1) + \mathcal{L}_{3d} = 0 \tag{2.11}$$

which after rearrangement using (B.2) solves as

$$\begin{aligned}\mathcal{L}_{3d} &= \frac{1}{1+\mathrm{e}^{2\mathrm{i}\pi d}}\left(-\frac{1+\mathrm{e}^{2\mathrm{i}\pi d}}{2d+1} + 2\mathrm{e}^{\mathrm{i}\pi d}B(d+1, d+1)\right) \\ &= \frac{B(d+1, d+1)}{\cos(\pi d)} - \frac{1}{2d+1}.\end{aligned} \tag{2.12}$$

By identity (B.8) and double applications of (B.15) and (B.13),

$$\cos(\pi d) = \frac{\sin(2\pi d)}{2\sin(\pi d)} = \frac{\Gamma(1-d)\Gamma(d+1)}{\Gamma(1-2d)\Gamma(2d+1)} \tag{2.13}$$

and also, by the second equality of (B.14) and (B.13),

$$B(d+1,d+1) = \frac{\Gamma(d+1)^2}{(2d+1)\Gamma(2d+1)}. \tag{2.14}$$

After substitution from (2.13) and (2.14) and simplification, (2.12) is seen to match the formula in (2.7). ∎

Proof of Theorem 2.2 Immediate on substituting (2.7) into (2.4) and simplifying. ∎

An alternative derivation of formula (2.6) is given in Chapter 9, using the harmonizable representation of the process; see Theorem **9.2**. Another version of the formula is

$$\Upsilon_d = \frac{\Gamma(d+1)^2}{\Gamma(2d+2)\cos \pi d}. \tag{2.15}$$

The proof that (2.15) matches (2.6) is a simple matter of applying (2.13) in reverse.

A useful identity, defined for any pair of time intervals $s_1 > t_1$ and $s_2 > t_2$, is

$$\begin{aligned}(X(s_1) - X(t_1))(X(s_2) - X(t_2)) \\ = \tfrac{1}{2}\big((X(s_1) - X(t_2))^2 + (X(s_2) - X(t_1))^2 \\ - (X(t_2) - X(t_1))^2 - (X(s_2) - X(s_1))^2\big).\end{aligned} \tag{2.16}$$

With Theorem **2.1** this allows the covariance of any pair of process increments to be calculated. Setting $t_1 = t_2 = 0$ in (2.16), and also $s_1 = s$ and $s_2 = t$, gives

$$\mathrm{E}(X(s)X(t)) = \tfrac{1}{2}\sigma_u^2\Upsilon_d\left(s^{2d+1} + t^{2d+1} - |s-t|^{2d+1}\right) \tag{2.17}$$

which with $d = 0$ reduces to the familiar case of regular Brownian motion, having $\mathrm{E}(X(s)X(t)) = \sigma_u^2 \min\{t, s\}$. For another example, set $t_1 = s_2 = t$, $s_1 = t + \delta$, and $t_2 = t - \delta$ in (2.16) to give the covariance of a pair of adjacent nonoverlapping increments,

$$\mathrm{E}((X(t+\delta) - X(t))((X(t) - X(t-\delta)) = \sigma_u^2\Upsilon_d(2^{2d} - 1)\delta^{2d+1}. \tag{2.18}$$

This is 0 if $d = 0$ and otherwise has the sign of d.

It is reasonable to ask how a stochastic process such as (2.1) can be understood to connect with the infinitely remote past. To elucidate this question imagine a version of (2.1) in which the Brownian motion driving the process has a finitely remote starting point, say $-N$ for some $N \in \mathbb{N}$. That is, define a process X^N by

$$X^N(t) = \int_0^t (t-\xi)^d \mathrm{d}U(\xi) + \int_{-N}^0 ((t-\xi)^d - (-\xi)^d)\mathrm{d}U(\xi), \; t \in [0,1]. \tag{2.19}$$

A minor amendment of Theorem **2.1** now gives the following.

2.4 Corollary $\mathrm{E}(X^N(t)^2) = \sigma_u^2 \Upsilon_d^N t^{2d+1}$ where

$$\Upsilon_d^N = \frac{1}{2d+1} + \int_0^N \left((1+\tau)^d - \tau^d\right)^2 \mathrm{d}\tau. \quad \square \tag{2.20}$$

It is easy to verify that for $t \in (0,1]$,

$$\begin{aligned} \mathrm{E}\left(X(t) - X^N(t)\right)^2 &= \sigma_u^2 t^{2d+1}\left(\Upsilon_d - \Upsilon_d^N\right) \\ &= \sigma_u^2 t^{2d+1} \int_N^\infty \left((1+\tau)^d - (\tau)^d\right)^2 \mathrm{d}\tau \\ &= O(N^{2d-1}) \end{aligned} \tag{2.21}$$

as $N \to \infty$. In other words, subject to the condition $d < \frac{1}{2}$, X exists as the mean square limit of a sequence of finite-lag processes X^N. This decomposition plays a major role in the treatment of the functional central limit theorem in Chapter 3.

2.3 The Linear Structure

The next objective is to show that when the shock sequence in (1.1) satisfies either Assumption **1.1** or Assumption **1.2**, the normalized partial sum of a fractional moving average that is covariance stationary and not overdifferenced converges to a limit having the variance function specified in Theorem **2.1**. The weak convergence proof itself is given in Chapter 3. The remainder of this preliminary chapter explores some essential properties of the objects under study and develops techniques of analysis.

With x_i given by (1.1) let

$$S_n = \sum_{i=1}^{n} x_i. \tag{2.22}$$

The representation in (1.3) suggests that this sum would be Gaussian in the limit if supplied with an appropriate normalization, although also showing that this normalization could not be $n^{-1/2}$. However, the framework of (2.22) is not ideal for calculating moments. S_n is a sum of sums and the essential trick, underlying all that follows, is to aggregate the components in a different order. Thus,

$$\begin{aligned} S_n &= \sum_{i=1}^{n}\sum_{j=0}^{\infty} b_j u_{i-j} \\ &= (b_0 u_n + b_1 u_{n-1} + \cdots) + (b_0 u_{n-1} + b_1 u_{n-2} + \cdots) \\ &\qquad + \cdots + (b_0 u_1 + b_1 u_0 + \cdots) \\ &= b_0 u_n + (b_0 + b_1) u_{n-1} + \cdots + (b_0 + \cdots + b_{n-1}) u_1 \\ &\qquad + (b_1 + \cdots + b_n) u_0 + (b_2 + \cdots + b_{n+1}) u_{-1} + \cdots \\ &= \sum_{i=-\infty}^{n} a_{ni} u_i \end{aligned} \tag{2.23}$$

where

$$a_{ni} = \sum_{j=\max\{0,1-i\}}^{n-i} b_j. \tag{2.24}$$

In this representation of S_n the terms are either independent or, at worst, weakly dependent, but they form a nonstationary sequence in view of (2.24) and are infinite in number.

More generally, the notation to be used in the sequel is

$$S_{[ns]} - S_{[nt]} = \sum_{i=-\infty}^{[ns]} a_{ni}(s,t) u_i \tag{2.25}$$

for any $0 \le t < s \le 1$, where

$$a_{ni}(s,t) = \sum_{j=\max\{0,[nt]-i+1\}}^{[ns]-i} b_j. \tag{2.26}$$

In particular, note that $a_{ni}(t,t) = 0$ for any i, representing an empty sum, so that $S_{[ns]} - S_{[nt]} = 0$ if $s = t$. Also, $a_{n[ns]}(s,t) = b_0 = 1$ when n is large enough that $[ns] > [nt]$. Where convenient, a_{ni} as in (2.24) will continue to be used as shorthand for the case $a_{ni}(1,0)$.

Assuming for simplicity that the coefficients b_j vary like (1.2) for all j and not just in the limit, consider how the sequence of weights $a_{ni}(s,t)$ in (2.26) varies in the contrasting cases $d > 0$ and $d < 0$. When $d > 0$, $a_{ni}(s,t)$ would take its maximum at $i = [nt] + 1$, at which point it has $[ns] - [nt]$ terms and diverges at the rate n^d. As i increases, the number of terms in $a_{ni}(s,t)$ decreases down to the single term $b_0 = 1$ at the point $i = [ns]$. Moving in the other direction, with $i \le [nt]$, the terms of the sum in (2.26) are $b_{1-i}, \ldots, b_{[ns]-i}$. While there are $[ns]$ of these terms and when $d > 0$ they are not summable, for a given fixed n their sum is tending to zero, with $a_{ni} = O(|i|^{d-1})$ as $i \to -\infty$.

In the antipersistent case with $d < 0$, $a_{n[ns]}(s,t) = b_0 = 1$ but the b_j are negative for $j > 0$, with $\sum_{j=0}^{\infty} b_j = 0$. Letting i decrease from $[ns]$ down to 1 the sum accumulates negative terms and so declines towards zero. There is a discontinuity in the function at the point $i = [nt]$ where it jumps to approximately -1 when n is large, being the sum of the first $[ns]$ coefficients starting from $j = 1$. Thereafter it forms an increasing (absolutely decreasing) negative sequence, tending to 0 with $a_{ni} = O(|i|^{d-1})$ as before.

These are the same patterns that characterize the weight functions of the limit process in (2.1). Viewed as partial sums, the integrated fractional processes detailed in §1.2 have fBm as their suitably normalized limit distribution. The process whose differences were described as antipersistent in §1.4 is the partial sum whose limit process has positive but declining weights when $\xi > 0$ and negative weights when $\xi < 0$. Being independent of initial conditions and so reverting on average to the centre line following excursions, this process exhibits negative autocorrelation according to (2.18) and can be crudely described as 'mean reverting'. These properties are further discussed in §4.4 on page 83, which looks at the case $x = y$ in the bivariate analysis.

With this framework established, let

$$\kappa(n) = n^{d+1/2}L(n) \tag{2.27}$$

and hence define the normalized partial sum process $X_n : [0,1] \mapsto \mathbb{R}$ where

$$X_n(t) = \frac{S_{[nt]}}{\kappa(n)} = \frac{1}{\kappa(n)} \sum_{i=-\infty}^{[nt]} a_{ni}(t,0)u_i. \tag{2.28}$$

Following (2.25) the variation over an interval $(t, s]$ has the form

$$X_n(s) - X_n(t) = \frac{1}{\kappa(n)} \sum_{i=-\infty}^{[ns]} a_{ni}(s,t)u_i. \tag{2.29}$$

The process defined by (2.28) is a step function, being constant except for jumps at the points where $t = [nt]/n$ and taking its value at the terminal points of the jumps. While not a continuous function of time it is right-continuous, with every point having a limit point to its left. Such processes are commonly referred to by the French acronym 'càdlàg'. If the process has the unit interval as domain, as is commonly the case, the space of càdlàg processes is denoted $D_{[0,1]}$, of which the space of continuous processes, $C_{[0,1]}$, forms a subset.

A technical detail that can be overlooked in the present context, but matters for convergence proofs, is that $D_{[0,1]}$ is customarily endowed with the Skorokhod J$_1$ topology (see [71], [72]). Distances between elements x and y of $C_{[0,1]}$ are defined by the so-called uniform metric, $d_U(x,y) = \sup_t |x(t) - y(t)|$, but this measure has undesirable properties in the presence of discontinuities. Without going into too much detail,[1] the Skorokhod distance between càdlàg processes x and y, denoted $d_S(x,y)$, is the smallest number by choice of homeomorphism $\lambda : [0,1] \mapsto [0,1]$ such that both $\sup_t |x(t) - y(\lambda(t))| \le d_S(x,y)$ and $\sup_t |\lambda(t) - t| \le d_S(x,y)$. Thus, functions are close in the Skorokhod topology if their discontinuities are close in time as well as in magnitude. In practice, distances are assigned by a separable complete metric embodying the topology, such as that constructed by Billingsley ([6], [7]).

2.4 Limiting Forms

Allowing the general formulation in (1.2) for the lag coefficients, the decomposition in (2.26) exhibits the following tendencies as n increases. In this context, note that 0^{d-1} stands in for b_0 in the orders of magnitude calculations relating to $\{b_j\}_{j=0}^{\infty}$ and is assigned the value 1.

2.5 Theorem Let $a_{ni}(s,t)$ be defined by (2.26) for $0 \le t < s \le 1$ where the b_j satisfy (1.2) for $|d| < \frac{1}{2}$ and $\sum_{j=0}^{\infty} b_j = 0$ if $d < 0$. For $t < x < s$,

$$a_{n[nx]}(s,t) \sim (n(s-x))^d L(n(s-x)) \tag{2.30}$$

[1] A full account can be found in SLT Chapter 30.

and for $-\infty < x \le t$,

$$a_{n[nx]}(s,t) \sim \left((n(s-x))^d - (n(t-x))^d\right) L(n(s-x)) \tag{2.31}$$

as $n \to \infty$.

Proof First, let $0 < d < \frac{1}{2}$. For positive integer m and slowly varying function L,

$$\int_0^m y^{d-1} L(y) \mathrm{d}y = \sum_{j=0}^{m-1} \int_j^{j+1} y^{d-1} L(y) \mathrm{d}y$$

where for large enough j,

$$(j+1)^{d-1} L(j+1) \le \int_j^{j+1} y^{d-1} L(y) \mathrm{d}y \le j^{d-1} L(j) \tag{2.32}$$

and hence for large enough m,

$$\sum_{j=1}^{m} j^{d-1} L(j) \le \int_0^m y^{d-1} L(y) \mathrm{d}y \le \sum_{j=0}^{m-1} j^{d-1} L(j). \tag{2.33}$$

For the case $t < x < s$, set $m = [ns] - [nx]$ in (2.33). It follows by (2.26) and (1.2) that

$$a_{n[nx]}(s,t) = \sum_{j=0}^{[ns]-[nx]} b_j \sim d \int_0^{n(s-x)} y^{d-1} L(y) \mathrm{d}y. \tag{2.34}$$

By Theorem **A.1**(i) of Appendix A, the integral in (2.34) has solution (2.30) as $n \to \infty$. If $-\infty < x \le t$ then

$$\begin{aligned} a_{n[nx]}(s,t) &= \sum_{j=[nt]+1-[nx]}^{[ns]-[nx]} b_j \\ &\sim d \int_0^{n(s-x)} y^{d-1} L(y) \mathrm{d}y - d \int_0^{n(t-x)} y^{d-1} L(y) \mathrm{d}y \end{aligned} \tag{2.35}$$

with solution matching (2.31).

Next, consider the case $-\frac{1}{2} < d < 0$ and $\sum_{j=0}^{\infty} b_j = 0$. By the same reasoning, if $t < x < s$ then

$$a_{n[nx]}(s,t) = \sum_{j=0}^{[ns]-[nx]} b_j = -\sum_{j=[ns]+1-[nx]}^{\infty} b_j$$
$$\sim -d \int_{n(s-x)}^{\infty} y^{d-1} L(y) dy$$

and the solution is given in this case by Theorem **A.1**(ii). If $x \le t$, similarly,

$$\begin{aligned} a_{n[nx]}(s,t) &= \sum_{j=[nt]+1-[nx]}^{[ns]-[nx]} b_j \\ &= \sum_{j=[nt]+1-[nx]}^{\infty} b_j - \sum_{j=[ns]+1-[nx]}^{\infty} b_j \\ &\sim d \int_{n(t-x)}^{\infty} y^{d-1} L(y) dy - d \int_{n(s-x)}^{\infty} y^{d-1} L(y) dy. \end{aligned} \quad \blacksquare$$

The next result applies these tendencies to show the implication of having Formula (1.2) constrain the moving average coefficients at long range.

2.6 Theorem If $|d| < \frac{1}{2}$,

$$\sum_{i=-\infty}^{[ns]} a_{ni}(s,t)^2 \sim \Upsilon_d (n(s-t))^{2d+1} L(n(s-t))^2. \tag{2.36}$$

Proof Define sums M_{1n} and M_{2n} by

$$\sum_{i=-\infty}^{[ns]} a_{ni}(s,t)^2 = \sum_{i=[nt]+1}^{[ns]} a_{ni}(s,t)^2 + \sum_{i=-\infty}^{[nt]} a_{ni}(s,t)^2 = M_{1n} + M_{2n}. \tag{2.37}$$

For fixed values of s and t (not indicated explicitly) let A_{ni} denote the approximator in (2.30) for $i = [nx]$ and let

$$g_{ni} = \frac{|a_{ni}^2(s,t) - A_{ni}^2|}{A_{ni}^2}. \tag{2.38}$$

Summing over the indices $i = [nt]+1, \ldots, [ns]$ that define M_{1n},

$$\left| \frac{\sum_i a_{ni}^2(s,t)}{\sum_i A_{ni}^2} - 1 \right| \le \frac{\sum_i g_{ni} A_{ni}^2}{\sum_i A_{ni}^2} \le \max_{[nt]<i\le[ns]} g_{ni} \to 0 \tag{2.39}$$

as $n \to \infty$, by the modulus inequality and Theorem **2.5**. Approximating sum by integral analogously to the argument leading to relation (2.34) and applying Theorem **A.1**(i), with $d > -\frac{1}{2}$ the convergence in (2.39) implies

$$M_{1n} \sim \sum_{i=[nt]+1}^{[ns]} A_{ni}^2 \sim \int_0^{n(s-t)} y^{2d} L(y)^2 \mathrm{d}y \sim \frac{(n(s-t))^{2d+1}}{2d+1} L(n(s-t))^2. \quad (2.40)$$

For the case $-\infty < i \leq [nt]$, substitute the approximator from (2.31) into (2.38) and construct the counterpart of (2.40) for M_{2n} to get

$$\begin{aligned} M_{2n} &\sim \int_{-\infty}^{nt} \left((ns-y)^d - (nt-y)^d\right)^2 L(ns-y)^2 \mathrm{d}y \\ &\sim (n(s-t))^{2d+1} L(n(s-t))^2 \int_0^{\infty} \left((1+\tau)^d - \tau^d\right)^2 \mathrm{d}\tau. \end{aligned} \quad (2.41)$$

The second asymptotic equivalence makes the change of variable $\tau = (nt - y)/(ns - nt)$ and uses the fact that $L(n(s-t)x)/L(n(s-t)) \to 1$ as $n \to \infty$ for both $x = \tau$ and $x = 1 + \tau$. ∎

The limiting variance of an increment of the process X_n is going to be a central feature of the asymptotic analysis and the following are immediate consequences of (2.25) and Theorem **2.6**.

2.7 Corollary Under Assumption **1.1** and with $|d| < \frac{1}{2}$,

$$\mathrm{E}\left(\sum_{i=[nt]+1}^{[ns]} x_i\right)^2 \sim \sigma_u^2 \Upsilon_d (n(s-t))^{2d+1} L(n(s-t))^2. \quad \square \quad (2.42)$$

2.8 Corollary For X_n defined in (2.28) with $t \in [0,1]$ and $\delta \in (0, 1-t]$, under Assumption **1.1** and $|d| < \frac{1}{2}$,

$$\lim_{n\to\infty} \mathrm{E}\left(X_n(t+\delta) - X_n(t)\right)^2 = \sigma_u^2 \Upsilon_d \delta^{2d+1}. \quad (2.43)$$

Proof Definition (2.28) implies

$$\mathrm{E}\left(X_n(t+\delta) - X_n(t)\right)^2 = \frac{\sigma_u^2}{\kappa(n)^2} \sum_{i=-\infty}^{[n(t+\delta)]} a_{ni}(t+\delta, t)^2. \quad (2.44)$$

Substitute into (2.44) from (2.36) and simplify. ∎

These calculations go some way to explaining the behaviour of the process at the boundary points $-\frac{1}{2}$ and $+\frac{1}{2}$. In the nonstationary case $d = \frac{1}{2}$ the integral in (2.41) diverges, showing that the remote data points are so influential that the partial sum process eventually blows up and cannot approach the limit distribution (2.1). The rate at which the tail component in (2.21) disappears with N is an indicator of what happens as the boundary is approached. In the case $d = -\frac{1}{2}$ the integral in (2.40) diverges. The reasoning in this case is a little more subtle, but the limit in (2.43) gives the clue. The constant Υ_d might be removed by choice of normalization and it is not so much that this quantity diverges with $d = -\frac{1}{2}$, as the fact that δ^{2d+1} does not approach 0 as $\delta \to 0$. This points to the failure of a.s. continuity.

2.5 The Multivariate Model

Extending the fractional model to describe the interactions of two or more related processes is very largely a matter of setting up appropriate notation. Let $\boldsymbol{\Delta}(B)$ denote a diagonal $m \times m$ matrix polynomial whose diagonal elements are of the form $b_k(B) = \sum_{j=0}^{\infty} b_{kj}B^j$ for $k = 1, \dots, m$, where B denotes the backshift operator and $b_{kj} \sim d_k j^{d_k - 1} L_k(j)$, with $|d_k| < \frac{1}{2}$ for each k. The parameters $d_1, \dots, d_m$ are assumed to be in order of magnitude with $d_1 \leq \dots \leq d_m$ and the functions $L_1(j), \dots, L_m(j)$ are either slowly varying or constant at infinity. A m-vector of fractional processes is then represented by

$$\boldsymbol{x}_i = \boldsymbol{\Delta}(B)\boldsymbol{u}_i \;\; (m \times 1). \tag{2.45}$$

Define $m \times m$ diagonal matrices

$$\boldsymbol{D}_n = \mathrm{diag}(n^{d_1+1/2}L_1(n), \dots, n^{d_m+1/2}L_m(n)) \tag{2.46}$$

and

$$\boldsymbol{A}_{ni}(s,t) = \mathrm{diag}(a_{1ni}(s,t), \dots, a_{mni}(s,t)) \tag{2.47}$$

where $a_{kni}(s,t)$ is the case of (2.26) with $d = d_k$ and so define the vector $\boldsymbol{X}_n$ $(m \times 1)$ in terms of its increments, by

$$\boldsymbol{X}_n(s) - \boldsymbol{X}_n(t) = \boldsymbol{D}_n^{-1} \sum_{i=[nt]+1}^{[ns]} \boldsymbol{x}_i = \boldsymbol{D}_n^{-1} \sum_{i=-\infty}^{[ns]} \boldsymbol{A}_{ni}(s,t)\boldsymbol{u}_i \tag{2.48}$$

for $0 \leq t < s \leq 1$.

Assume that the elements of the m-vector shock process $\{\boldsymbol{u}_i, -\infty < i < \infty\}$ individually satisfy the conditions of Assumption **1.1** and have the contemporaneous covariance matrix

$$\boldsymbol{\Sigma}_u = \{\sigma_{kl}\} = \mathrm{E}(\boldsymbol{u}_i\boldsymbol{u}_i') \ (m \times m). \tag{2.49}$$

Also define the matrix $\boldsymbol{\Upsilon} = \{\Upsilon_{kl}\}$ $(m \times m)$ where

$$\Upsilon_{kl} = \frac{1}{d_k + d_l + 1} + \int_0^\infty ((1+\tau)^{d_k} - \tau^{d_k})((1+\tau)^{d_l} - \tau^{d_l})\mathrm{d}\tau. \tag{2.50}$$

The Hadamard product of the matrices $\boldsymbol{\Sigma}_u$ and $\boldsymbol{\Upsilon}$, written $\boldsymbol{\Sigma}_u \odot \boldsymbol{\Upsilon}$, is the $m \times m$ matrix having elements $\sigma_{kl}\Upsilon_{kl}$ for $k, l = 1, \ldots, m$. Finally, for $0 < \delta \leq 1$ let

$$\boldsymbol{K}(\delta) = \mathrm{diag}(\delta^{d_1+1/2}, \ldots, \delta^{d_m+1/2}).$$

This setup permits the following generalization of Corollary **2.8**.

2.9 Theorem If Assumption **1.1** holds for each element of $\boldsymbol{u}_i$,

$$\lim_{n\to\infty} \mathrm{E}(\boldsymbol{X}_n(t+\delta) - \boldsymbol{X}_n(t))(\boldsymbol{X}_n(t+\delta) - \boldsymbol{X}_n(t))' = \boldsymbol{K}(\delta)(\boldsymbol{\Sigma}_u \odot \boldsymbol{\Upsilon})\boldsymbol{K}(\delta) \tag{2.51}$$

for $0 < \delta \leq 1$ and $0 \leq t < 1 - \delta$.

Proof Modify the proof of Theorem **2.6** as follows. Replace the term $a_{ni}(s,t)^2$ in (2.37) by $a_{kni}(s,t)a_{lni}(s,t)$ and replace A_{ni}^2 by the product $A_{kni}A_{lni}$ where these factors represent the approximator functions (2.30) and (2.31) evaluated at parameters d_k and d_l. The result that

$$\Upsilon_{kl} = \lim_{n\to\infty} \frac{1}{(n\delta)^{d_k+d_l+1}L_k(n)L_l(n)} \sum_{i=-\infty}^{[n(t+\delta)]} a_{kni}(t+\delta, t)a_{lni}(t+\delta, t) \tag{2.52}$$

for each pair k, l and any choice of t and δ is then a straightforward extension. The modifications of the asymptotic equivalences in (2.40) and (2.41) are direct, with $d_k + d_l$ replacing $2d$ and the squared terms replaced by products in the cases $k \neq l$.

Given (2.52), the argument of Corollary **2.8** can be applied to each element of the expected outer product of (2.51) with one modification, that according to (2.48) the $(k,l)^{\text{th}}$ divisor has the form $n^{d_k+d_l+1}L_k(n)L_l(n)$ in place of

$\kappa(n)^2$ as appears in equation (2.44). The conclusion is therefore that, for each pair (k, l),

$$\mathrm{E}(X_{kn}(t+\delta) - X_{kn}(t))(X_{ln}(t+\delta) - X_{ln}(t)) \to \sigma_{kl}\Upsilon_{kl}\delta^{d_k+d_l+1}. \tag{2.53}$$

These limits constitute the elements of the limit matrix in (2.51). ∎

Another way to express this limit is to use the fact that if $\boldsymbol{B} = \mathrm{diag}(\boldsymbol{b})$ where $\boldsymbol{b}$ is $m \times 1$ then $\boldsymbol{B\Sigma}_u\boldsymbol{B} = \boldsymbol{\Sigma}_u \odot \boldsymbol{bb}'$. Therefore, (2.51) can also be written as

$$\lim_{n\to\infty} \sum_{i=-\infty}^{[n(t+\delta)]} \boldsymbol{A}_{ni}(t+\delta, t)\boldsymbol{D}_n^{-1}\boldsymbol{\Sigma}_u\boldsymbol{D}_n^{-1}\boldsymbol{A}_{ni}(t+\delta, t)$$
$$= \boldsymbol{\Sigma}_u \odot (\boldsymbol{K}(\delta)\boldsymbol{\Upsilon}\boldsymbol{K}(\delta)) = \boldsymbol{K}(\delta)(\boldsymbol{\Sigma}_u \odot \boldsymbol{\Upsilon})\boldsymbol{K}(\delta).$$

$\boldsymbol{\Upsilon}$ is symmetric and positive semidefinite by construction and according to the Schur product theorem the Hadamard product of positive (semi)definite matrices is positive (semi)definite. $\boldsymbol{\Upsilon}$ is singular if two or more of the fractional parameters are equal, in particular equal to 0, but $\boldsymbol{\Sigma}_u \odot \boldsymbol{\Upsilon}$ should have the same rank as $\boldsymbol{\Sigma}_u$ in that case. If the first m_1 rows and columns of $\boldsymbol{\Upsilon}$ are equal to one another then all the elements of $\boldsymbol{\Upsilon}_{11}$ $(m_1 \times m_1)$ take the same value, say $v_{11} > 0$, and the submatrix $\boldsymbol{\Upsilon}_{12}$ $(m_1 \times m_2)$ also has rank 1. However, letting $\boldsymbol{v}_{12} > \boldsymbol{0}$ be the m_2-vector of the common column elements of $\boldsymbol{\Upsilon}_{12}$, with the matching partition of $\boldsymbol{\Sigma}_u$ the Hadamard product would have the form

$$\boldsymbol{\Sigma}_u \odot \boldsymbol{\Upsilon} = \begin{bmatrix} v_{11}\boldsymbol{\Sigma}_{u11} & \boldsymbol{\Sigma}_{u12}\mathrm{diag}(\boldsymbol{v}_{12}) \\ \mathrm{diag}(\boldsymbol{v}_{12})\boldsymbol{\Sigma}_{u12}' & \boldsymbol{\Sigma}_{u22}\odot\boldsymbol{\Upsilon}_{22} \end{bmatrix}.$$

For example, in the case $d_k = 0$ for $k = 1, \ldots, m_1$, $v_{11} = 1$ and $\boldsymbol{v}_{12}$ has elements $1/(1+d_l)$ for $l = m_1 + 1, \ldots, m$.

A closed form for (2.50) is given in Chapter 4, as Theorem **4.6**. The context of this latter result is a bivariate analysis, but a covariance is necessarily a pairwise construction. Adapted to the present notation, the formula in (4.36) takes the form

$$\Upsilon_{kl} = \frac{\Gamma(d_k+1)\Gamma(d_l+1)\cos(\pi(d_k-d_l)/2)}{\Gamma(d_k+d_l+2)\cos(\pi(d_k+d_l)/2)}. \tag{2.54}$$

As required, $\Upsilon_{kl} = \Upsilon_{lk}$ and the formula collapses to Υ_d, as in (2.15), on setting $d_k = d_l = d$. A direct derivation based on the harmonizable representation of the processes is given in Theorem **9.3**.

It is tempting to ask whether formula (2.16) has a generalization to allow the calculation of covariances for increments over differing time intervals. Unequivocally the answer to this question is no, since covariances can shed no light on the time-ordering of events. This is clear from Theorem **9.3**, since by construction the harmonizable representation of the processes can contain no information about temporal orderings. To illustrate the position, try modifying (2.16) for a sequence pair (X_k, X_l), replacing the squares of increments with contemporaneous products. For simplicity's sake set $t_1 = t_2 = 0$ assuming $X_k(0) = X_l(0) = 0$, also putting $s_1 = s$ and $s_2 = t$. This gives the equality

$$X_k(s)X_l(s) + X_k(t)X_l(t) - (X_k(s) - X_k(t))(X_l(s) - X_l(t))$$
$$= X_k(s)X_l(t) + X_k(t)X_l(s). \tag{2.55}$$

The expectation of the left-hand side of (2.55) can be calculated by (2.53), but it is not possible to determine either $\mathrm{E}(X_k(s)X_l(t))$ or $\mathrm{E}(X_k(t)X_l(s))$ individually, only their sum.

2.6 Shock Dependence

Replacing Assumption **1.1** by Assumption **1.2** often results in no change to asymptotic behaviour beyond the replacement of σ_u^2 by ω_u^2 in formulae, but this can be quite tedious to demonstrate. As an example consider Corollary **2.8**, which is an easy consequence of Theorem **2.6**. Its generalization, given here initially for the case $t = 0$ and $\delta = 1$, depends on a blocking argument. Recall the definition of $\kappa(n)$ in (2.27).

2.10 Theorem Under Assumption **1.2** and $|d| < \frac{1}{2}$,

$$\lim_{n\to\infty} \mathrm{E}\left(\frac{1}{\kappa(n)}\sum_{i=1}^{n} x_i\right)^2 = \omega_u^2 \Upsilon_d.$$

Proof Write a_{ni} as in (2.24) to stand for $a_{ni}(1,0)$. Choose an increasing integer sequence $\{B_n\}$ such that $B_n \to \infty$ but $B_n/n \to 0$ and let $r_n = [n/B_n]$. Define, for $j = -\infty, \ldots, r_n$,

$$S_{nj} = \frac{1}{B_n^{1/2}} \sum_{i=(j-1)B_n+1}^{jB_n} u_i \tag{2.56}$$

and also

$$S^*_{nj} = \sum_{i=(j-1)B_n+1}^{jB_n} \frac{a_{ni} - a_{n,jB_n}}{B_n^{3/2} g_{nj}} u_i \tag{2.57}$$

where $g_{nj} > 0$ is an array to be chosen. Decompose the sum using (2.23), as

$$\frac{1}{\kappa(n)} \sum_{i=1}^{n} x_i = \frac{1}{\kappa(n)} \sum_{i=-\infty}^{n} a_{ni} u_i = A_{1n} + A_{2n} + A_{3n} \tag{2.58}$$

where

$$A_{1n} = \frac{B_n^{1/2}}{\kappa(n)} \sum_{j=-\infty}^{r_n} a_{n,jB_n} S_{nj}$$

$$A_{2n} = \frac{B_n^{3/2}}{\kappa(n)} \sum_{j=-\infty}^{r_n} g_{nj} S^*_{nj}$$

$$A_{3n} = \frac{1}{\kappa(n)} \sum_{i=r_nB_n+1}^{n} a_{ni} u_i.$$

The object of the proof is to show that

$$\lim_{n\to\infty} \mathrm{E}(A_{1n}^2) = \omega_u^2 \Upsilon_d \tag{2.59}$$

whereas $\mathrm{E}(A_{2n}^2)$ and $\mathrm{E}(A_{3n}^2)$ and hence also the expected cross-products are of small order in the limit.

Since the autocovariances are summable by Assumption **1.2**(a),

$$\sup_{-\infty<j\le r_n} |\mathrm{E}(S_{nj}^2) - \omega_u^2| \to 0 \text{ as } n \to \infty. \tag{2.60}$$

The block products $\mathrm{E}(S_{nj}S_{n,j-m})$ for $m \ge 1$ define B_n^2 pairings of coordinates having date separations $B_n m + p$ that vary from $B_n(m-1)+1$ with $p = 1 - B_n$ to $B_n(m+1) - 1$ with $p = B_n - 1$. Since $B_n - |p|$ of the pairs have separation $B_n m + p$ for each p, Assumption **1.2**(a) implies that

$$\begin{aligned} |\mathrm{E}(S_{nj}S_{n,j-m})| &\le \frac{1}{B_n} \sum_{p=1-B_n}^{B_n-1} (B_n - |p|)|\gamma_u(B_n m + p)| \\ &= O(m^{-1-\delta} B_n^{-\delta}) \end{aligned} \tag{2.61}$$

and hence that

$$\sup_{-\infty<j\le r_n} \sum_{m=1}^{\infty} |\mathrm{E}(S_{nj}S_{n,j-m})| = O(B_n^{-\delta}). \tag{2.62}$$

The next step is to define

$$w_{nj} = \frac{B_n^{1/2} a_{n,jB_n}}{\kappa(n)}, \; -\infty < j \le r_n \tag{2.63}$$

and so to write

$$\mathrm{E}(A_{1n}^2) = T_{1n} + 2T_{2n} \tag{2.64}$$

where

$$T_{1n} = \sum_{j=-\infty}^{r_n} w_{nj}^2 \mathrm{E}(S_{nj}^2)$$

and

$$T_{2n} = \sum_{j=-\infty}^{r_n-1} \sum_{m=1}^{r_n-j} w_{nj} w_{n,j+m} \mathrm{E}(S_{nj}S_{n,j+m}).$$

The analysis of Theorem **2.6** with n replaced by r_n gives the result

$$\sum_{j=-\infty}^{r_n} a_{r_n j}^2 \sim \Upsilon_d r_n^{2d+1} L(r_n)^2. \tag{2.65}$$

Then, since $(n - B_n j)^d \sim B_n^d(r_n - j))^d$ for $0 \le j \le r_n$, $(-B_n j)^d = B_n^d(-j))^d$ for $j < 0$, and $L(r_n)/L(n) \to 1$, replacing each $a_{r_n j}$ by a_{n,jB_n} in (2.65) gives according to Theorem **2.5**

$$B_n \sum_{j=-\infty}^{r_n} a_{n,jB_n}^2 \sim \Upsilon_d \kappa(n)^2.$$

This implies according to (2.63) that $\sum_{j=-\infty}^{r_n} w_{nj}^2 \to \Upsilon_d$ and hence by (2.60) that

$$|T_{1n} - \omega_u^2 \Upsilon_d| \le \sup_{-\infty<j\le r_n} |\mathrm{E}(S_{nj}^2 - \omega_u^2)| \sum_{j=-\infty}^{r_n} w_{nj}^2 + \omega_u^2 \left| \sum_{j=-\infty}^{r_n} w_{nj}^2 - \Upsilon_d \right|$$

$$\to 0 \tag{2.66}$$

as $n \to \infty$. Also, since $\sum_{j=-\infty}^{r_n-1} |w_{nj} w_{n,j+m}| \le \Upsilon_d$ for $1 \le m \le r_n - j$ when n is large enough, (2.62) implies that

$$|T_{2n}| = O(B_n^{-\delta}). \tag{2.67}$$

The implication of (2.64), (2.66), and (2.67), together with (2.36), is that (2.59) is confirmed.

Next consider A_{2n}, for which constants g_{nj} must be chosen. In view of formula (2.26) with $t = 0$ and $s = 1$ the term $a_{ni} - a_{n,B_nj}$ contains the sum of the $B_nj - i$ successive coefficients $b_{n-B_nj+1}, \ldots, b_{n-i}$. If the sequence $\{|b_j|\}_{j=1}^{\infty}$ is monotone decreasing then $|a_{ni} - a_{n,B_nj}| \leq (B_nj - i)|b_{n-B_nj+1}|$ and in this case let the choice in (2.57) be

$$g_{nj} = |b_{n-B_nj+1}| \tag{2.68}$$

so that $|a_{ni} - a_{n,B_nj}| \leq B_n g_{nj}$ for $B_n(j-1) < i \leq B_nj$. More generally, set $g_{nj} = \max_{m \geq n-B_nj+1} |b_m|$. Given (1.2), the resulting shift is by at most a finite number of steps and leaves the asymptotic argument unaffected. With this setup, the weights appearing in (2.57) are bounded by $B_n^{-1/2}$ and by arguments paralleling those for S_{nj},

$$\sup_{-\infty < j \leq r_n} \mathrm{E}(S_{nj}^{*2}) = O(1) \tag{2.69}$$

and

$$\sup_{-\infty < j \leq r_n} \sum_{m=1}^{\infty} |\mathrm{E}(S_{nj}^* S_{n,j-m}^*)| = O(B_n^{-\delta}). \tag{2.70}$$

The next step is to define

$$w_{nj}^* = \frac{B_n^{3/2} g_{nj}}{\kappa(n)}, \quad -\infty < j \leq r_n. \tag{2.71}$$

In view of the facts that $|b_{n-B_nj+1}| \sim (n - B_nj)^{d-1} L(n - B_nj)$ by (1.2), that $n \sim r_n B_n$, and that $2d - 2 < -1$, the choice of g_{nj} in (2.68) implies thanks to summability that

$$\sum_{j=-\infty}^{r_n} B_n^3 g_{nj}^2 \sim B_n^{2d+1} \sum_{j=-\infty}^{r_n} (r_n - j)^{2d-2} L(n - B_nj)^2 = O(B_n^{2d+1})$$

and so, according to (2.71),

$$\sum_{j=-\infty}^{r_n} w_{nj}^{*2} = O(r_n^{-2d-1}). \tag{2.72}$$

From (2.69), (2.70), and (2.72), the conclusion is that

$$\begin{aligned} \mathrm{E}(A_{2n}^2) &= \sum_{j=-\infty}^{r_n} w_{nj}^{*2}\mathrm{E}(S_{nj}^{*2}) + 2\sum_{j=-\infty}^{r_n-1}\sum_{m=1}^{r_n-j} w_{nj}^{*}w_{n,j+m}^{*}\mathrm{E}(S_{nj}^{*}S_{n,j+m}^{*}) \\ &= O(r_n^{-2d-1}) \end{aligned} \tag{2.73}$$

which is of small order since $d > -\frac{1}{2}$.

Finally, A_{3n} contains at most B_n terms and in view of (2.30) and Assumption **1.2**(a), $\mathrm{E}(A_{3n}^2) = O(r_n^{-2d-1})$. The proof is completed by noting that $\mathrm{E}(A_{1n}A_{2n})$, $\mathrm{E}(A_{1n}A_{3n})$, and $\mathrm{E}(A_{2n}A_{3n})$ are likewise of small order by the Cauchy–Schwarz inequality. ∎

For clarity of exposition this result has been set out for the case of $a_{ni}(1, 0)$, but the following extension is immediate.

2.11 Corollary Under Assumption **1.2** and $|d| < \frac{1}{2}$,

$$\lim_{n\to\infty} \mathrm{E}\left(\frac{1}{\kappa(n)}\sum_{i=[nt]+1}^{[ns]} x_i^2\right) = \omega_u^2\Upsilon_d(s-t)^{2d+1}.$$

Proof In Equation (2.58), a_{ni} is replaced by $a_{ni}(s, t)$, n by $[ns]$, so that r_n becomes $[[ns]/B_n]$. Wherever Υ_d appears it is replaced by $\Upsilon_d(s - t)^{2d+1}$. With these substitutions the proof of Theorem **2.10** is replicated almost unchanged. ∎

The next corollary of Theorem **2.10** extends the result to the multivariate case, in the manner of Theorem **2.9**. To replace the contemporaneous covariance matrix $\boldsymbol{\Sigma}_u$ defined in (2.49), the long-run covariance matrix is defined as

$$\boldsymbol{\Omega}_u = \{\omega_{kl}\} = \lim_{n\to\infty}\frac{1}{n}\sum_{i=1}^{n}\sum_{j=1}^{n}\mathrm{E}(\boldsymbol{u}_i\boldsymbol{u}_j') \ (m\times m) \tag{2.74}$$

and is assumed to be finite, generalizing Assumption **1.2**.

2.12 Corollary If Assumption **1.2** holds for each element of $\boldsymbol{u}_i$ in (2.48),

$$\lim_{n\to\infty}\mathrm{E}(\boldsymbol{X}_n(t+\delta) - \boldsymbol{X}_n(t))(\boldsymbol{X}_n(t+\delta) - \boldsymbol{X}_n(t))' = \boldsymbol{K}(\delta)(\boldsymbol{\Omega}_u \odot \boldsymbol{\Upsilon})\boldsymbol{K}(\delta). \tag{2.75}$$

Proof This is by extension of the arguments of Theorems **2.9** and **2.10** and Corollary **2.11**. In place of A_{1n}, A_{2n}, and A_{3n} in (2.58), define in the obvious manner the pairs $A_{1kn}, A_{1ln}, A_{2kn}, A_{2ln}$, and A_{3kn}, A_{3ln}. In (2.60), replace $\mathrm{E}(S_{nj}^2) - \omega_u^2$ with $\mathrm{E}(S_{knj}S_{lnj}) - \omega_{kl}$ and in (2.61), $\mathrm{E}(S_{nj}S_{n,j-m})$ with $\mathrm{E}(S_{knj}S_{ln,j-m})$ with the starred sums redefined in the same manner. In (2.64), $A_{1kn}A_{1ln}$ replaces A_{1n}^2 and $a_{kni}a_{lni}$ replaces a_{ni}^2, with the corresponding substitutions in (2.73). With these substitutions, the logic of the proof of Theorem **2.10** holds for each of the diagonal and off-diagonal elements of the matrix, with $\omega_{kl}\Upsilon_{kl}\delta^{d_k+d_l+1}$ replacing $\omega_u^2\Upsilon_d\delta^{2d+1}$. ∎

Chapter 3
The FCLT for Fractional Processes

Proofs of functional weak convergence of fractional partial sum processes were given originally by Yu. Davydov ([21]) and Murad Taqqu ([77]). Other contributions to this literature include [29] and more recently [80] and the relevant chapters of [28] and [5]. FCLTs for the Type II form of the limit process are given in [73], [40], and [49]. The material of the present chapter and specifically of §3.5 is based on joint work with Robert de Jong in [23] and [17], much expanded with the benefit of invaluable commentary in [42] in particular.

What distinguishes the present approach is that it deals with the Type I limit process and in §3.5 provides a version for general nonparametric dependence of the shocks as specified in Assumption **1.2**(b). The other results cited above all depend in one way or another on a linear short-memory structure based on independent shocks. Assumption **1.1** is also invoked here initially, for simplicity and so that a complete proof can be given without too much reliance on cited material. Theorem **3.18** applies a FCLT for dependent processes proved in SLT, to verify the Gaussian limit.

3.1 The Main Result

Let X_n be as defined in (2.28) with x_i given by (1.1) and (1.2) with $|d| < \frac{1}{2}$ and u_i satisfying Assumption **1.1**. Given the results of Chapter 2, a functional central limit theorem can be proved establishing the weak convergence of X_n to the fBm X defined in (2.1), with parameter d. The following extra assumption for the shock process in relation to d is also made, imposing a further moment restriction when $d < 0$ (the antipersistent case).

3.1 Assumption $\{u_i\}_{i=-\infty}^{\infty}$ is L_r-bounded for $r \geq \max\{2, 1/(\frac{1}{2} + d)\}$. □

3.2 Theorem If $|d| < \frac{1}{2}$ and Assumptions **1.1** and **3.1** hold, $X_n \to_d X$. □

Recall that X_n is an element of the càdlàg space $D_{[0,1]}$ equipped with the Skorokhod topology, as defined on page 23, while the limit process X is almost

Asymptotics for Fractional Processes. James Davidson, Oxford University Press. © James Davidson (2025).
DOI: 10.1093/9780198955207.003.0003

surely an element of $C_{[0,1]}$. The weak convergence indicated by the symbol '$\to_d$' is to be understood as a twofold phenomenon, in which the so-called finite-dimensional distributions of process increments tend to Gaussianity, and in addition the sequence of distributions indexed by sample size n is uniformly tight, such that the limit distribution is confined to $C_{[0,1]}$ with probability 1. This is explained in more detail in §3.4 and more about the associated distribution theory can be found in the chapters of Part VI of SLT. That the limit process has the covariance function of X is already known, given the agreement between Corollary **2.8** and Theorem **2.1**.

The proof of Theorem **3.2** is shown by means of a series of lemmas. As a preliminary, it is illuminating to write an increment of the limit process as the sum of three Brownian functionals. For a constant $N \in \mathbb{N}$, further decomposing (2.3) yields

$$X(s) - X(t) = \int_t^s (s-\xi)^d \mathrm{d}U(\xi) + \int_{-N}^t \left((s-\xi)^d - (t-\xi)^d\right) \mathrm{d}U(\xi)$$
$$+ \int_{-\infty}^{-N} \left((s-\xi)^d - (t-\xi)^d\right) \mathrm{d}U(\xi). \tag{3.1}$$

The first two terms of this decomposition constitute the increment $X^N(s) - X^N(t)$ according to the definition in (2.19). According to Corollary **2.4**, under the assumption on d, by taking N large enough the final term of (3.1) can be treated as negligible in L_2 norm, as shown in (2.21).

According to (2.29) the finite-n counterparts of these limit expressions are

$$X_n(s) - X_n(t) = \left(\sum_{i=[nt]+1}^{[ns]} + \sum_{i=1-nN}^{[nt]} + \sum_{i=-\infty}^{-nN}\right) \frac{a_{ni}(s,t)}{\kappa(n)} u_i$$
$$= R_{1n}(s,t) + R_{2n}(s,t) + R_{3n}(s,t). \tag{3.2}$$

The objective is to show that the increment $X^N(s) - X^N(t)$ represents the weak limit of the first two terms of (3.2). Let the sample counterpart be denoted by

$$X_n^N(t) = R_{1n}(t,0) + R_{2n}(t,0) \tag{3.3}$$

so that $R_{3n}(t,0) = X_n(t) - X_n^N(t)$. The status of this latter term is established similarly to Corollary **2.4**.

3.3 Lemma Under Assumption **1.1** and with $|d| < \frac{1}{2}$, $\lim_{n\to\infty} \mathrm{E}(R_{3n}(s,t)^2) = O(N^{2d-1})$ as $N \to \infty$.

Proof Following the approach of Theorem **2.6**, similarly to (2.41) and (2.21),

$$\begin{aligned}\mathrm{E}(R_{3n}(s,t))^2 &= \frac{\sigma_u^2}{\kappa(n)^2}\sum_{i=-\infty}^{-nN} a_{ni}^2(s,t) \\ &\sim \sigma_u^2(s-t)^{2d+1}\int_N^\infty ((1+\tau)^d - \tau^d)^2 \mathrm{d}\tau. \end{aligned} \tag{3.4}$$

The conclusion follows by (2.21). ∎

This result distinguishes the two modes of convergence as n and N respectively increase, which are to be thought of as strictly sequential. While it could be proved that $\mathrm{E}(R_{3n}(1,0))^2 \to 0$ as $N \to \infty$ for any finite value of n, it is only necessary to show this for the limit case. In the usual way with the functional convergence of partial sums, increasing n represents the compression of a growing set of observations into a finite interval. For any given N the discrete collection of points approaches a continuum as the gaps separating them shrink like $1/n$, leading to the asymptotic equivalence indicated in (3.4). By contrast, the divergence of N represents the inclusion of more remote lags into the sum. The domain of the function $R_{1n}(s,t) + R_{2n}(s,t)$ grows with N although more slowly than the number $nN + [ns]$ of corresponding data points.

Theorem **2.6** leads to the conclusions that under Assumption **1.1**,

$$\mathrm{E}(R_{1n}(s,t))^2 = \frac{\sigma_u^2}{\kappa(n)^2}\sum_{i=[nt]+1}^{[ns]} a_{ni}^2(s,t) \to \frac{\sigma_u^2(s-t)^{2d+1}}{2d+1} \tag{3.5}$$

and

$$\begin{aligned}\mathrm{E}(R_{2n}(s,t))^2 &= \frac{\sigma_u^2}{\kappa(n)^2}\sum_{i=1-nN}^{[nt]} a_{ni}^2(s,t) \\ &\to \sigma_u^2(s-t)^{2d+1}\int_0^N ((1+\tau)^d - \tau^d)^2 \mathrm{d}\tau \end{aligned} \tag{3.6}$$

as $n \to \infty$. Putting these limits together, it is possible to write

$$\mathrm{E}(X^N(s) - X^N(t))^2 = \sigma_u^2 \Upsilon_d^N (s-t)^{2d+1} \tag{3.7}$$

where Υ_d^N is defined in (2.20) and $\Upsilon_d - \Upsilon_d^N = O(N^{2d-1})$ according to (2.21). The procedure is to derive the limiting distributions of the increments $X_n^N(s) - X_n^N(t) = R_{1n}(s,t) + R_{2n}(s,t)$ where the approximation to (3.2) is controlled by the choice of N and can be made as close as desired.

3.2 Finite-dimensional Distributions

For the results of this section, Assumption **3.1** is not required.

3.4 Lemma If $|d| < \frac{1}{2}$ and Assumption **1.1** holds,

$$R_{1n}(s,t) + R_{2n}(s,t) \xrightarrow{\mathrm{d}} \mathrm{N}\left(0, \sigma_u^2 \Upsilon_d^N (s-t)^{2d+1}\right)$$

for any s and t with $0 \le t < s \le 1$.

Proof The terms $R_{1n}(s,t)$ and $R_{2n}(s,t)$ are shown to have Gaussian limits individually and since these pairs have no time periods in common, they are independent of each other by assumption so that their limiting sum is also Gaussian. The limiting variance follows directly from (3.5) and (3.6). The terms of the sums $R_{1n}(s,t)$ and $R_{2n}(s,t)$ are also independent under Assumption **1.1**, but are heterogeneously distributed with moving average weights $a_{ni}(s,t)/\kappa(n)$. The Lindeberg condition is therefore sufficient for the CLT to hold.[1]

For notational clarity the proof is given for the case $t = 0$ and $s = 1$ although the generalization is direct. For $R_{1n}(1,0)$, recalling (2.24), the Lindeberg condition specifies that for any $\varepsilon > 0$,

$$\lim_{n\to\infty} \frac{1}{\kappa(n)^2} \sum_{i=1}^{n} a_{ni}^2 \mathrm{E}\left(u_i^2 1_{\{|a_{ni}u_i|/\kappa(n) > \varepsilon\}}\right) = 0. \tag{3.8}$$

To check this condition, consider

$$\frac{1}{\kappa(n)^2} \sum_{i=1}^{n} a_{ni}^2 \mathrm{E}\left(u_i^2 1_{\{|a_{ni}u_i|/\kappa(n) > \varepsilon\}}\right)$$
$$\le \max_{1\le i\le n} \mathrm{E}\left(u_i^2 1_{\{|a_{ni}u_i|/\kappa(n) > \varepsilon\}}\right) \frac{1}{\kappa(n)^2} \sum_{i=1}^{n} a_{ni}^2. \tag{3.9}$$

[1] See SLT Theorem 24.6.

It follows by Theorem **2.6** and Corollary **2.8** that $\kappa(n)^{-2}\sum_{i=1}^{n} a_{ni}^2 = O(1)$. Therefore, to bound the majorant of (3.9) apply Theorem **A.5** in Appendix A with $\eta = \varepsilon\kappa(n)/|a_{ni}|$. There are two cases to be considered. In the case $d > 0$, Theorem **2.5** gives $\max_{1\le i\le n}|a_{ni}|/\kappa(n) = O(n^{-1/2})$, not overlooking that in ratios of this sort the slowly varying components, if any, cancel in the limit. According to Theorem **A.5**, the bound on (3.9) therefore has the form

$$\max_{1\le i\le n} \mathrm{E}\left(u_i^2 1_{\{|u_i|>\varepsilon\kappa(n)/|a_{ni}|\}}\right) = o\left(\max_{1\le i\le n}(|a_{ni}|/\kappa(n))^{r-2}\right) = o(n^{1-r/2}) \qquad (3.10)$$

which vanishes when $r \ge 2$. If $d < 0$, Theorem **2.5** says that $\max_{1\le i\le n}|a_{ni}| = 1$, and so $\max_{1\le i\le n}|a_{ni}|/\kappa(n) = O(n^{-d-1/2})$. In this case Theorem **A.5** gives

$$\max_{1\le i\le n} \mathrm{E}(u_i^2 1_{\{|u_i|>\varepsilon\kappa(n)/|a_{ni}|\}}) = o(n^{(d+1/2)(2-r)}). \qquad (3.11)$$

Therefore, provided $d > -\frac{1}{2}$ the Lindeberg condition is satisfied in either case when $r \ge 2$, although when $d < 0$ the convergence is correspondingly slower.

In the case of $R_{2n}(1,0)$, the Lindeberg condition assumes the form

$$\lim_{n\to\infty}\frac{1}{\kappa(n)^2}\sum_{i=1-nN}^{0} a_{ni}^2\mathrm{E}(u_i^2 1_{\{|a_{ni}u_i|/\kappa(n)>\varepsilon\}}) = 0 \qquad (3.12)$$

and Theorem **2.5** and Corollary **2.8** give

$$\frac{|a_{ni}|}{\kappa(n)} \sim \frac{|(n-i)^d - (-i)^d|}{n^{d+1/2}} \qquad (3.13)$$

which is decreasing in $-i$ for both $d > 0$ and $d < 0$. A convenient way to handle the Lindeberg condition test is to break up the terms of the sum into N blocks of length n, itemized by $k = 0, \ldots, N-1$. Since they are independent of each other by assumption, proving Gaussianity for each achieves this for the aggregate.

For the case $k = 0$, the first n terms of $R_{2n}(1,0)$ (counting backwards from zero) contribute the sum

$$\frac{1}{\kappa(n)^2}\sum_{i=1-n}^{0} a_{ni}^2\mathrm{E}(u_i^2 1_{\{|a_{ni}u_i|/\kappa(n)>\varepsilon\}})$$

$$\le \max_{1-n\le i\le 0}\mathrm{E}(u_i^2 1_{\{|a_{ni}u_i|/\kappa(n)>\varepsilon\}}\sum_{i=1-n}^{0}\frac{a_{ni}^2}{\kappa(n)^2}$$

where (3.13) implies

$$\frac{1}{\kappa(n)^2}\sum_{i=1-n}^{0} a_{ni}^2 \sim \int_0^1 ((\tau+1)^d - \tau^d)^2 d\tau.$$

Similarly to (3.10) and (3.11), the maximum of $|a_{ni}|$ in (3.13) over $1-n \le i \le 0$ is $O(n^d)$ when $d > 0$ but $O(1)$ when $d < 0$, so Theorem **A.5** gives

$$\max_{1-n\le i\le 0} \mathrm{E}(u_i^2 1_{\{|u_i|>\varepsilon\kappa(n)/|a_{ni}|\}}) = \begin{cases} o(n^{(1-r/2)}), & d > 0 \\ o(n^{(d+1/2)(2-r)}), & d < 0. \end{cases} \tag{3.14}$$

For the cases $k \ge 1$, the maximum of $|a_{ni}|$ over the range $1 - n(k+1) \le i \le -nk$ with n large enough is at $i = -nk$ for d of either sign. According to (3.13), $|a_{n,-nk}| \sim n^d|(k+1)^d - k^d|$ and so according to Theorem **A.5**,

$$\begin{aligned}\max_{-n(k+1)<i\le -nk} \mathrm{E}(u_i^2 1_{\{|u_i|/>\varepsilon\kappa(n)/|a_{ni}|\}}) &= o\left(\left(\frac{n^{d+1/2}}{n^d|k+1)^d - k^d|}\right)^{2-r}\right) \\ &= o(n^{1-r/2}).\end{aligned}$$

Since $((k+1)^d - k^d)^2$ is a summable sequence when $|d| < \frac{1}{2}$, the conclusion according to (3.13) is that for d of either sign,

$$\begin{aligned}&\sum_{k=1}^{N-1}\sum_{i=1-n(k+1)}^{-nk} \frac{a_{ni}^2}{\kappa(n)^2}\mathrm{E}\left(u_i^2 1_{\{|a_{ni}u_i|/\kappa(n)>\varepsilon\}}\right) \\ &\quad\le n\sum_{k=1}^{N-1} \max_{-n(k+1)<i\le -nk} \frac{a_{ni}^2}{\kappa(n)^2}\mathrm{E}\left(u_i^2 1_{\{|u_i|>\varepsilon\kappa(n)/|a_{ni}|\}}\right) \\ &\quad\ll \sum_{k=1}^{N-1}((k+1)^d - k^d)^2 \max_{-n(k+1)<i\le -nk} \mathrm{E}\left(u_i^2 1_{\{|u_i|>\varepsilon\kappa(n)/|a_{ni}|\}}\right) \\ &\quad= o(n^{1-r/2}).\end{aligned} \tag{3.15}$$

The result for $R_{2n}(1,0)$ is confirmed by (3.14) and (3.15).

This has proved the theorem for the case $t = 0$, $s = 1$. Replacing a_{ni} by $a_{ni}(s,t)$, and changing the sum limits from $1,\dots,n$ to $[nt]+1,\dots,[ns]$ and from $1-nN,\dots,0$ to $1-nN,\dots,[nt]$ extends the results to $R_{1n}(s,t)$ and $R_{2n}(s,t)$ with straightforward (albeit unattractively complicated) modifications of the formulae, completing the proof as a whole. ∎

3.3 Uniform Boundedness and Uniform Integrability

The hypothesized limit process (2.1) is a.s. continuous. It remains to show that the sequence of distributions defined for $\{X_n^N, n \in \mathbb{N}\}$, where X_n^N is defined in (3.3), has a limit that is a.s. continuous likewise. The key concept of uniform tightness of the sequence is reviewed in §3.4, but first it has to be established that process increments of a fixed width δ are uniformly bounded and uniformly square-integrable. The tightness analysis then considers what happens as δ shrinks to zero.

Consider an increment $X_n^N(t+\delta) - X_n^N(t)$ for $\delta > 0$ and $0 \le t \le 1-\delta$. For $s \in (t, t+\delta]$ define

$$T_n(s,t) = X_n^N(s) - X_n^N(t) = R_{1n}(s,t) + R_{2n}(s,t) \tag{3.16}$$

and decompose the variance correspondingly. Under Assumption **1.1**, $\mathrm{E}(T_n(t+\delta,t)^2) = \sigma_u^2 v_n^2(t,\delta)$ where

$$\begin{aligned} v_n^2(t,\delta) &= \frac{1}{\kappa(n)^2}\left(\sum_{i=[nt]+1}^{[n(t+\delta)]} + \sum_{i=1-nN}^{[nt]}\right) a_{ni}^2(t+\delta,t) \\ &= v_{1n}^2(t,\delta) + v_{2n}^2(t,\delta). \end{aligned} \tag{3.17}$$

The variances of $R_{1n}(t+\delta,t)$ and $R_{2n}(t+\delta,t)$ are $\sigma_u^2 v_{1n}^2(t,\delta)$ and $\sigma_u^2 v_{2n}^2(t,\delta)$ respectively and $v_n^2(t,\delta) = O(1)$ in view of (2.36) and (2.27).

It is helpful to set up a compact notation for the objects under study. For $\delta > 0$ and $0 \le t \le 1-\delta$, define

$$\widetilde{T}_n(t,\delta) = \sup_{t\le s\le t+\delta} \frac{|T_n(s,t)|}{v_n(t,\delta)}. \tag{3.18}$$

The notations $\widetilde{R}_{1n}(t,\delta)$ and $\widetilde{R}_{2n}(t,\delta)$ are defined likewise for the processes in question, here normalized by $v_{1n}(t,\delta)$ or $v_{2n}(t,\delta)$ as the case may be.

The squared supremum of the absolute value is the same thing as the supremum of the square. For given t and δ, the uniform square-integrability of (3.18) is the condition that for some $n_0 < \infty$,

$$\sup_{n>n_0} \mathrm{E}\left(\widetilde{T}_n^2(t,\delta) 1_{\{\widetilde{T}_n(t,\delta)>B\}}\right) \to 0 \text{ as } B \to \infty. \tag{3.19}$$

In fact, a stronger condition can be shown specifying the rate of convergence of the expectations to zero and this extension will be needed in the sequel.

While $n_0 = 1$ may generally be a reasonable assumption, it is possible for the required conditions to be imposed on the limiting case but left unspecified in finite samples.

Uniform integrability cannot hold without uniform boundedness in probability and the following result is therefore informative, for while Assumption **3.1** is shown in Theorem **3.8** to be sufficient for uniform tightness, the fact of its necessity in the antipersistent case is an important feature of the fractional asymptotics. The following necessity proof adapts a counter-example in [81], cited by [42] in commenting on [17].

3.5 Theorem Under Assumption **1.1**, for $\delta > 0$ and $0 \le t \le 1 - \delta$ the collection $\{\widetilde{T}_n^2(t,\delta), n \in \mathbb{N}\}$ is uniformly bounded in probability if and only if Assumption **3.1** holds.

Proof If u_i is L_r-bounded then according to Lemma **A.4**, $P(|u_i| > \eta) = O(\eta^{-r}\log(\eta)^{-1-\mu})$ as $\eta \to \infty$ for $\mu > 0$. If the u_i are independently and identically distributed then for any $\varepsilon > 0$,

$$\begin{aligned}
P\left(\sup_{t\le s\le t+\delta}\frac{|u_{[ns]}|}{\kappa(n)} \le \varepsilon\right) &= P\left(\bigcap_{i=[nt]}^{[n(t+\delta)]}\{|u_i| \le \varepsilon\kappa(n)\}\right)\\
&= (1 - P(|u_i| > \varepsilon\kappa(n)))^{[n(t+\delta)]-[nt]+1}\\
&= O\left(\left(1 - \frac{1}{\varepsilon^r\kappa(n)^r\log(\varepsilon\kappa(n))^{1+\mu}}\right)^{n\delta}\right)\\
&= O\left(\exp\left\{-\frac{n\kappa(n)^{-r}}{\varepsilon^r\log(\varepsilon\kappa(n))^{1+\mu}}\right\}^{\delta}\right). \qquad (3.20)
\end{aligned}$$

If $n\kappa(n)^{-r} \to \infty$ then this probability converges to zero as $n \to \infty$ for all $\varepsilon > 0$ and the supremum over $[nt] \le i \le [n(t+\delta)]$ of $|u_i|/\kappa(n)$ accordingly diverges with probability 1.

According to (2.27) the condition $n\kappa(n)^{-r} = O(1)$ holds by Assumption **3.1**. The proof is completed by showing that this condition is both necessary and sufficient for $\sup_{t\le s\le t+\delta}|T_n(s,t)|$ to be bounded in probability. According to (2.26), the partial sum from $[nt]$ to $[ns]$ has the form

$$\begin{aligned}
T_n(s,t) &= \sum_{i=1-nN}^{[ns]}\frac{a_{ni}(s,t)}{\kappa(n)}u_i\\
&= \frac{b_0}{\kappa(n)}u_{[ns]} + \frac{b_0+b_1}{\kappa(n)}u_{[ns]-1} + \frac{b_0+b_1+b_2}{\kappa(n)}u_{[ns]-2} + \cdots \qquad (3.21)
\end{aligned}$$

where $b_0 = 1$ and if $d < 0$ the lag coefficients in (3.21) are converging to 0 as in (1.19). To show necessity, rearrange (3.21) by pulling out the leading term, taking absolute values, applying the triangle inequality, and then taking the sup of both sides, to give

$$\sup_{t \le s \le t+\delta} \frac{|u_{[ns]}|}{\kappa(n)} \le \sup_{t \le s \le t+\delta} |T_n(s,t)| + \sup_{t \le s \le t+\delta} \left| \sum_{i=1-nN}^{[ns]-1} \frac{a_{ni}(s,t)}{\kappa(n)} u_i \right|. \tag{3.22}$$

Noting that the second majorant term does not contain $u_{[ns]}$ and that $\nu_n(t,\delta) = O(1)$, the divergence of the minorant of (3.22) with probability 1 implies the divergence of $\widetilde{T}_n^2(t,\delta)$ likewise.

To show sufficiency, consider relation (3.20) for the case with $t = -N$ and δ set to $N + 1$. Under Assumption **3.1**, this has the implication that as $\varepsilon \to \infty$,

$$-\log P\left(\sup_{-N \le s \le 1} \frac{|u_{[ns]}|}{\kappa(n)} \le \varepsilon \right) = O(\varepsilon^{-r} \log(\varepsilon\kappa(n))^{-1-\mu}). \tag{3.23}$$

The probability itself therefore tends to 1. By definition the bound in question applies to every $u_i/\kappa(n)$ in the sum (3.21) and it follows that $|T_n(s,t)| = O_p(1)$ under Assumptions **1.1** and **3.1**. This is true for any s and in particular for $\sup_{t \le s \le t+\delta} |T_n(s,t)|$. The same order of magnitude extends to (3.18) and, as noted, the supremum of the square is the square of the supped absolute value. ∎

To see the implications of this result for (3.19), rearrange the Chebyshev inequality to get

$$\begin{aligned} \varepsilon^2 \left(1 - P(\widetilde{T}_n(t,\delta) \le \varepsilon)\right) &< \mathrm{E}(\widetilde{T}_n^2(t,\delta)) \\ &= \mathrm{E}\left(\widetilde{T}_n^2(t,\delta) 1_{\{\widetilde{T}_n(t,\delta) < \varepsilon\}}\right) + \mathrm{E}\left(\widetilde{T}_n^2(t,\delta) 1_{\{\widetilde{T}_n(t,\delta) \ge \varepsilon\}}\right). \end{aligned} \tag{3.24}$$

Unless Assumption **3.1** holds, (3.22) implies that if $d < 0$, with large enough n the probability in the minorant of (3.24) must shrink so far that $\mathrm{E}\left(\widetilde{T}_n^2(t,\delta)\right) \ge \varepsilon^2$, which means that the second term in the right-hand member cannot vanish. As $n \to \infty$ this remains true even as ε is taken arbitrarily large, so (3.19) is ruled out.

The natural route to demonstrating that (3.19) holds is by verifying a uniform bound for squared partial sums, such as Theorem **A.8** in Appendix A. However, the argument is not straightforward due to the fact that the moving average weights appearing in $T_n(s,t)$ are not merely an array of constants but depend on s, over which the supremum is to be taken. Therefore the collection $\{\widetilde{T}_n^2(t,\delta), n \in \mathbb{N}\}$ does not meet the specified conditions of Theorem **A.8**.

What can be shown is that this sequence is dominated by sequences that do meet the conditions. The key feature of Theorem **A.8** in this context is that the constant weights c_{nj} are arbitrary and the Doob inequality continues to hold if the ordering of the weights is permuted, given that the sum of their squares is unchanged. The following lemma develops a roundabout application of the theorem to obtain the required result.

3.6 Lemma Under Assumptions **1.1** and **3.1**, for all $0 < \delta < 1$ and $t \in [0, 1-\delta]$ the collection $\{\widetilde{T}_n^2(t,\delta), n \in \mathbb{N}\}$ is uniformly integrable with

$$\mathrm{E}\left(\widetilde{T}_n^2(t,\delta)1_{\{\widetilde{T}_n(t,\delta)>B\}}\right) = o(B^{2-r}). \tag{3.25}$$

Proof Consider the sum (3.16). Taking absolute values and applying the triangle inequality, also noting $\nu_n(t,\delta) \ge \nu_{1n}(t,\delta)$ and $\nu_n(t,\delta) \ge \nu_{2n}(t,\delta)$, gives the relation

$$\frac{|T_n(s,t)|}{\nu_n(t,\delta)} \le \frac{|R_{1n}(s,t)|}{\nu_{1n}(t,\delta)} + \frac{|R_{2n}(s,t)|}{\nu_{2n}(t,\delta)}. \tag{3.26}$$

Taking the sup over s of each term of (3.26) and then squaring both sides leads to the inequality

$$\widetilde{T}_n^2(t,\delta) \le \left(\widetilde{R}_{1n}(t,\delta) + \widetilde{R}_{2n}(t,\delta)\right)^2 \tag{3.27}$$

which holds with probability 1. According to Theorems **A.7** and **A.6**, to prove the lemma it is sufficient to show that the collections $\{\widetilde{R}_{1n}^2(t,\delta), n \in \mathbb{N}\}$ and $\{\widetilde{R}_{2n}^2(t,\delta), n \in \mathbb{N}\}$ are uniformly integrable, with order of magnitude $o(B^{2-r})$ as $B \to \infty$ similarly to (3.25).

Consider the moving average coefficients of $R_{1n}(s,t)$. According to (2.26) these have the form $a_{ni}(s,t)/\kappa(n)$ where $a_{ni}(s,t) = b_0 + \cdots + b_k$ with $k = [ns] - i$. Here, i runs from $[nt]+1$ to $[ns]$ and hence k runs from $[ns]-[nt]-1$ down to 0 as i increases. The complete set of these coefficients, by descending order of i, is

$$\left\{\frac{b_0}{\kappa(n)}, \frac{b_0+b_1}{\kappa(n)}, \frac{b_0+b_1+b_2}{\kappa(n)}, \ldots, \frac{b_0+\cdots+b_{[n(t+\delta)]-[nt]-1}}{\kappa(n)}\right\}. \tag{3.28}$$

For any $s \in (t, t+\delta]$ the coefficients in $R_{1n}(s,t)$ belong to the set (3.28), being the first $[ns]-[nt]$ members of the indicated sequence, but in reverse order. For example, the coefficient multiplying $u_{[nt]+1}$ changes each time $[ns]$ increases, being the kth element of (3.28) for $k = 0, 1, 2, \ldots, [n(t+\delta)]-[nt]-1$ as $[ns]$ ranges from $[nt]+1$ up to $[n(t+\delta)]$. The fact that the successive partial sums as s increases are not a simple cumulation of terms is the problem with the application of Theorem **A.8**.

Therefore, consider for $s \in (t, t+\delta]$ the modified array

$$R^*_{1n}(s,t) = \frac{1}{\kappa(n)} \sum_{i=[nt]+1}^{[ns]} a_{ni}(t+\delta,t)u_i \tag{3.29}$$

in which the moving average coefficients depend on fixed t and δ, but not on s. Then, $R^*_{1n}(t+\delta,t) = R_{1n}(t+\delta,t)$ but for other values of s the coefficients in (3.29), while also drawn from the set (3.28) in decreasing order as i increases, start from the right-hand end according to (2.26). Define

$$c_{ni} = \frac{|a_{ni}(t+\delta,t)|}{\kappa(n)\nu_{1n}(t,\delta)}$$

so that $\sum_{i=[nt]+1}^{[n(t+\delta)]} c_{ni}^2 = 1$ by (3.17) and

$$\frac{R^*_{1n}(s,t)}{\nu_{1n}(t,\delta)} = \sum_{i=[nt]+1}^{[ns]} c_{ni}u_i. \tag{3.30}$$

This is the cumulation of the first $[ns] - [nt]$ terms of a sequence with fixed coefficients. It follows by Theorem **A.8** that, on the assumptions, the collection $\{\widetilde{R}^*_{1n}(t,\delta)^2, n \in \mathbb{N}\}$ is uniformly integrable with the rate of convergence in (3.25).

This property does not depend on the ordering of the moving average weights in the sum. Theorem **A.8** continues to apply under arbitrary permutations of the indices, with terms $c_{np(i)}u_i$ (say) replacing $c_{ni}u_i$ in (3.30). Since $\nu^2_{1n}(t,\delta)$ is the sum of all the squared elements of (3.28), these permutations leave it unchanged.

Letting $R^*_{1np}(s,t)$ denote the partial sum of the $c_{np(i)}u_i$, a permutation p^* can always be chosen such that, with probability 1,

$$\sup_{\{t\le s\le t+\delta\}} R_{1n}(s,t)^2 \le \sup_{\{t\le s\le t+\delta\}} R^*_{1np^*}(s,t)^2. \tag{3.31}$$

To see that such a p^* exists, note how the sup in the minorant of (3.31) is constructed. In effect, the sum of the terms $\{a_{ni}(s,t)u_i, i = [nt]+1,\dots,[ns]\}$ is formed for each s, and s is then chosen to maximize the square of this sum. By contrast, on the majorant side of (3.31) the partial sums are formed by picking the coefficients $c_{np^*(i)}$ in the indicated sequence from the set (3.28). This set includes all the terms appearing in the minorant and equality in (3.31) is always achieved by having the two sets of coefficients matching. Free choice of p^* never need do worse than this.

Let $\{p_n^*\}$ denote a sequence of permutations that satisfy inequality (3.31) for each n, and so form the collection $\{\widetilde{R}^*_{1np_n^*}(t,\delta)^2, n \in \mathbb{N}\}$. By construction, each element of this collection is drawn from a sequence having the property shown in (3.25) and hence shares that property. In view of (3.31) and Theorem **A.6** the collection $\{\widetilde{R}_{1n}(t,\delta)^2, n \in \mathbb{N}\}$ has the same property.

Next, consider $R_{2n}(s,t)$ for $t \le s \le t+\delta$. These sums all contain $[nt] + nN$ terms, but they differ because for each $[ns]$ the moving average coefficients are sums of $[ns]-[nt]$ terms according to (2.26). Thus, for each $i = 1-nN, \dots, [nt]$, $a_{ni}(s,t) = b_{[nt]+1-i} + \cdots + b_{[ns]-i}$. The second term of (3.17), $v_{2n}^2(t,\delta)$, is the sum of the coefficients $a_{ni}^2(t+\delta,t)/\kappa(n)^2$. The sum of the terms $a_{ni}^2(s,t)/\kappa(n)^2$ for a given s may similarly be written $v_{2n}^2(t, s-t)$. If the weights c_{ni} are defined for this s by the equality

$$\frac{R_{2n}(s,t)}{v_{2n}(t,\delta)} = \sum_{i=1-nN}^{[nt]} c_{ni}u_i \tag{3.32}$$

then for this same s,

$$\sum_{i=1-nN}^{[nt]} c_{ni}^2 = \frac{v_{2n}^2(t,s-t)}{v_{2n}^2(t,\delta)}. \tag{3.33}$$

If the ratio in (3.33) is $O(1)$ as $n \to \infty$, the sums (3.32) satisfy the conditions of Corollary **A.9**. This is true for any choice of s in the closed interval $[t, t+\delta]$ and hence, in particular, for the collection $\{\widetilde{R}_{2n}^2(t,\delta), n \in \mathbb{N}\}$.

It therefore remains to show that the ratio (3.33) is indeed $O(1)$. The sums $v_{2n}^2(t,s-t)$ and $v_{2n}^2(t,\delta)$ contain the same number of terms, respectively the terms $a_{ni}^2(s,t)$ and $a_{ni}^2(t+\delta,t)$ defined by (2.26) and with large enough n, $a_{ni}^2(s,t) \le a_{ni}^2(t+\delta,t)$ for every i. This is because the coefficients b_j have the sign of d at long range according to (1.2) and hence sums with more terms (larger s for given i) are absolutely larger. This is true even with $d < 0$, since $i \le [nt]$ and hence b_0 is always excluded. Thus, there exists $n_0 < \infty$ large enough that $v_{2n}^2(t,s-t) \le v_{2n}^2(t,\delta)$ for $n \ge n_0$. ∎

n_0 is left unspecified in this result only because (1.2) leaves the finite-order lag coefficients unspecified. In the case of (1.13), to take the leading example, the coefficients are a monotone sequence and n_0 can be set to 1. Note, however, that without Assumption **3.1**, with $d < 0$ the expectations in (A.13) are undefined in the limit as $n \to \infty$, as demonstrated by Theorem **3.5** and (3.24), so that the arguments based on Theorem **A.8** fail. The assumption is accordingly necessary.

3.4 Uniform Tightness

A tight distribution is one having the property that there exists a compact subset of the sample space having probability arbitrarily close to 1, ruling out the case of positive probability attaching to an infinite outcome. The chief concern is that a sequence of probability measures defined by increasing sample size might tend towards a violation of this condition. Uniform tightness is the property of a sequence of tight distributions, that tightness is preserved in the limit. A familiar counter-example is the sequence of uniform distributions on the intervals $[-n, n]$ of $\mathbb{R}$, which are well defined for every finite n but not in the limit.

The empirical process X_n defined by (2.28) is an element of the space $D_{[0,1]}$ of càdlàg functions on the unit interval, in other words, a step function. As the sample size n increases, the width of the steps shrinks like $1/n$ while the heights of the jumps are diminishing in line with the standard deviation of the increments, at the rate $n^{-1/2-d}$. Notwithstanding the Gaussianity at particular points, to match fBm in (2.1) the limit distribution of the process must be confined to the space $C_{[0,1]}$ with probability 1. This property of a sequence of random empirical processes is called stochastic equicontinuity. A formal definition and additional discussion of this concept can be found in, for example, §22.3 of SLT.

A step function can be transformed into a continuous function by the simple expedient of joining the vertices with straight lines. Hence, it is also possible to think in terms of a sequence of distributions with domain $C_{[0,1]}$ and the issue is then whether the limit process is also in $C_{[0,1]}$, with probability 1. If some lines do not shrink in length to zero in the limit but instead become vertical jumps as the step width goes to zero (and hence, diverge under normalization) the process is said to escape from $C_{[0,1]}$ in the limit. Uniform tightness fails if this occurrence has positive probability. A sequence with domain $D_{[0,1]}$ is accordingly said to be uniformly tight if the limit process is in $C_{[0,1]}$ almost surely, which is the property required in the present case.

To allow additional applications, specifically those arising in Chapter 6, in the following results the domain of the empirical processes is taken to be an interval $[L, U]$ that is finite and includes the origin, but otherwise is unspecified. For the present purpose of proving Theorem **3.2** the bounds are set to $L = 0$ and $U = 1$. For a random process $Y_n \in D_{[L,U]}$, the so-called modulus of continuity is defined as

$$w_n(\delta) = \sup_{t\in[L,U]} \sup_{\{s:|s-t|<\delta\}} |Y_n(s) - Y_n(t)|. \tag{3.34}$$

If $w_n(\delta) \to 0$ as $\delta \to 0$, the function Y_n must be continuous in its argument. If stochastic equicontinuity holds, the probability of this event can be made as close to one as desired by taking n large enough and uniform tightness of the sequence can be defined in these terms. Sufficient conditions are proved as Theorem 30.19 of SLT, which is itself adapted from Theorem 15.5 of [6]. Adapted to the present applications and omitting some topological detail, this theorem can be cast in the following form.

3.7 Theorem Let $\{\mu_n\}$ denote a sequence of probability measures on $D_{[L,U]}$ where $U - L < \infty$. If for $n \geq n_0$ with $n_0 < \infty$,
(a) for $\eta > 0$ there exists $M < \infty$ such that $\mu_n(x : \sup_t |x(t)| > M) \leq \eta$
(b) for $\varepsilon > 0$ and $\eta > 0$ there exists $\delta > 0$ such that $\mu_n(w_n(\delta) \geq \varepsilon) \leq \eta$
then $\{\mu_n\}$ is uniformly tight and any cluster point has $\mu_n(C_{[L,U]}) = 1$. □

SLT Theorem 30.19 treats the case $L = 0$ and $U = 1$ and also replaces condition **3.7**(a) with the condition of boundedness in probability at a point of the interval, which is sufficient given condition **3.7**(b) although condition **3.7**(a) is convenient for the present case since it has already been shown to hold by Theorem **3.5**.

The next result draws inspiration from SLT Theorem 31.5, which is based in turn on Theorem 3.1 of [23]. Stochastic equicontinuity is shown by proving the validating condition **3.7**(b).

3.8 Theorem For a random process $Y_n \in D_{[L,U]}$ with $U - L < \infty$ and $\delta > 0$, let

$$\widetilde{T}_n(t,\delta) = \sup_{t \leq s \leq t+\delta} \frac{|Y_n(s) - Y_n(t)|}{\nu_n(t,\delta)}. \tag{3.35}$$

If $|d| < \frac{1}{2}$ and for each $t \in [L, U]$ and $n \geq n_0$ for $n_0 < \infty$,

(a) $\nu_n(t,\delta) = O(\delta^{\min\{1/2, d+1/2\}})$ as $\delta \to 0$

(b) $\mathrm{E}\left(\widetilde{T}_n^2(t,\delta) 1_{\{\widetilde{T}_n(t,\delta) > B\}}\right) = o(B^{2-r})$ for $r \geq \max\{2, 1/(d+\frac{1}{2})\}$

then for $w_n(\delta)$ defined in (3.34) and $\varepsilon > 0$,

$$\lim_{\delta \to 0} \limsup_{n \to \infty} P(w_n(\delta) \geq \varepsilon) = 0. \quad \square \tag{3.36}$$

The familiar P is here interpreted to denote μ_n when the argument is a random element bearing subscript n. In the case $Y_n = X_n^N$ defined in (3.3), which is the present application of the theorem, the function $\nu_n(t,\delta)$ appearing in definition (3.35) is defined by $\mathrm{E}(X_n^N(t+\delta) - X_n^N(t))^2 = \sigma_u^2 \nu_n(t,\delta)^2$ under Assumption **1.1**. However, this is not a condition of the theorem and under Assumption **1.2** σ_u^2 would be replaced by ω_u^2.

Proof of Theorem 3.8 For $\delta > 0$ consider the set of integers $J(\delta) = \{0, ..., [(U-L)/\delta] - 1\}$. The collection $\{t_j = L + j\delta : j \in J(\delta)\}$ contains equally spaced points of $[L, U]$ with $t_0 = L$ and $t_{[(U-L)/\delta]-1} > U - 2\delta$. For any pair $s, t \in [L, U]$ such that $|s - t| < \delta$, the largest $j \in J(\delta)$ such that both $t \geq t_j$ and $s \geq t_j$ has the property $t \leq t_j + 2\delta$ and $s \leq t_j + 2\delta$. In view of the relation $\{|x + y| > \varepsilon\} \subseteq \{|x| > \frac{1}{2}\varepsilon\} \cup \{|y| > \frac{1}{2}\varepsilon\}$ for variables x and y, subadditivity implies that for the t_j having this property in relation to the indicated pair (s, t),

$$P(w_n(\delta) > \varepsilon) = P\left(\sup_{t\in[L,U]} \sup_{\{s:|s-t|<\delta\}} |Y_n(s) - Y_n(t_j) + Y_n(t_j) - Y_n(t)| > \varepsilon \right)$$

$$\leq 2P\left(\max_{j\in J(\delta)} w_{nj}(\delta) > \tfrac{1}{2}\varepsilon \right) \tag{3.37}$$

where

$$w_{nj}(\delta) = \sup_{t_j\leq t\leq t_j+2\delta} |Y_n(t) - Y_n(t_j)|.$$

Noting that $w > M$ for $M > 0$ if and only if $w1_{\{w>M\}} > M$, subadditivity and then the Markov inequality also give

$$P\left(\max_{j\in J(\delta)} w_{nj}(\delta) > \tfrac{1}{2}\varepsilon \right) \leq \sum_{j\in J(\delta)} P\left(w_{nj}^2(\delta) > \tfrac{1}{4}\varepsilon^2 \right)$$

$$\leq \frac{4}{\varepsilon^2} \sum_{j\in J(\delta)} \mathrm{E}\left(w_{nj}^2(\delta) 1_{\{w_{nj}^2(\delta)>\varepsilon^2/4\}} \right). \tag{3.38}$$

For brevity, let $\bar{v}_{nj} = v_n(t_j, 2\delta)$ and define $\bar{w}_{nj} = w_{nj}(\delta)/\bar{v}_{nj}$, dependence on δ being understood. In view of condition (a), $\bar{v}_{nj}^2 = O((2\delta)^{\min\{1,2d+1\}})$ as $\delta \to 0$ and since the set $J(\delta)$ has $[(U-L)/\delta]$ elements, $\sum_{j\in J(\delta)} \bar{v}_{nj}^2 \ll 2(U-L)\delta^{\min\{0,2d\}}$. Hence,

$$\sum_{j\in J(\delta)} \mathrm{E}\left(w_{nj}^2(\delta) 1_{\{w_{nj}^2(\delta)>\varepsilon^2/4\}} \right) = \sum_{j\in J(\delta)} \bar{v}_{nj}^2 \mathrm{E}\left(\bar{w}_{nj}^2 1_{\{\bar{w}_{nj}^2>\varepsilon^2/4\bar{v}_{nj}^2\}} \right)$$

$$\ll 2(U-L)\delta^{\min\{0,2d\}} \max_{j\in J(\delta)} \mathrm{E}\left(\bar{w}_{nj}^2 1_{\{\bar{w}_{nj}^2>\varepsilon^2/4 \max_{k\in J(\delta)} \bar{v}_{nk}^2\}} \right). \tag{3.39}$$

Passing back up the chain of inequalities (3.37), (3.38), and (3.39), the probability in the minorant of (3.37) goes to zero as $\delta \to 0$ if the same is true of the majorant of (3.39). It follows that the convergence in (3.36) occurs if two conditions are met. The first is that

$$\lim_{\delta\to 0}\limsup_{n\to\infty}\sup_{t\in[0,1-\delta]} v_n^2(t,\delta) = 0 \tag{3.40}$$

so that the argument of the indicator in the majorant of (3.39) diverges. This follows directly from assumption (a) since $d > -\frac{1}{2}$. The second condition is that, for any $n \geq n_0$, the majorant of (3.39) vanishes as a result of (3.40). Note from (3.35) that $\bar{w}_{nj} = \widetilde{T}_n(t_j, 2\delta)$. Applying assumption (b), put $B = \varepsilon/2\max_{k\in J(\delta)}\bar{\nu}_{nk}$ and assumption (a) bounds the majorant of (3.39) by

$$\delta^{\min\{0,2d\}}\max_{j\in J(\delta)} \mathrm{E}\left(\bar{w}_{nj}^2 1_{\{|\bar{w}_{nj}|>\varepsilon/2\max_{k\in J(\delta)}\bar{\nu}_{nk}\}}\right)$$
$$= \begin{cases} o(\delta^{r/2-1}), & d \geq 0 \\ o(\delta^{2d-(1/2+d)(2-r)}), & d < 0 \end{cases} \tag{3.41}$$

as $\delta \to 0$. Since $2d - (\frac{1}{2} + d)(2 - r) \geq 0$ under the bound on r in assumption (b), the proof is complete. ∎

Proof of Theorem 3.2 It is sufficient to show that $X_n^N \to_{\mathrm{d}} X^N$ for $N \in \mathbb{N}$, where X_n^N is defined in (3.3). Since $X^N \to_{L_2} X$ as $N \to \infty$ by Lemma **3.3**, the limiting distribution of X_n^N may be made by choice of N as close in mean square as desired to the distribution of X.

Lemma **3.4** establishes the Gaussianity of X^N. Letting $\widetilde{T}_n(s,t)$ in (3.35) be defined by (3.18) and (3.16), apply Theorems **3.7** and **3.8** for the case $Y_n = X_n^N$ with $L = 0$ and $U = 1$. The fact that for $\eta > 0$ there exists $M < \infty$ such that $P(\widetilde{T}_n(0,1) > M) \leq \eta$ for $n \geq n_0$ and $n_0 < \infty$ follows from the sufficiency part of Theorem **3.5** and confirms condition (a) of Theorem **3.7**. Theorem **3.8**, whose assumptions are verified by (3.7) and Lemma **3.6**, in turn verifies condition (b) of Theorem **3.7**, so proving uniform tightness and the a.s. continuity of the limit distribution. Corollary **2.8** establishes the covariance structure, to complete the proof. ∎

3.5 Dependent Shocks

Proving the FCLT under the assumption of independent shocks highlights the special features of the problem while keeping complexities to a minimum, but the ambition must be to allow weak dependence. One reason would be to allow the moving average coefficients in (1.2) to be given a preferred functional form. The obvious example is the fractional integration formula (1.13), without which it is not possible to give the fractionally integrated process the neat representation

$$x_i = (1 - B)^{-d} u_i$$

from (1.8). This formulation might be regarded as losing little generality over (1.1)+(1.2) provided u_i is not required to be an independent process.

A feature of the results of this section is a dependence on material from SLT. These involve, among other things, the following new assumption.

3.9 Assumption For scale constants $\{c_{ni}, i = 1, \ldots, n\}$, there exists $\alpha \in (0, 1]$ such that with $B_n = [n^{1-\alpha}]$ and $r_n = [n/B_n]$ for $n \in \mathbb{N}$, the following conditions hold: if

$$M_{nj} = \max_{(j-1)B_n+1 \leq i \leq jB_n} |c_{ni}| \tag{3.42}$$

for $j = 1, \ldots, r_n$ and $M_{n,r_n+1} = \max_{r_n B_n+1 \leq i \leq n} |c_{ni}|$ then

$$\max_{1 \leq j \leq r_n+1} M_{nj} = o(B_n^{-1/2}) \tag{3.43}$$

and

$$\sum_{j=1}^{r_n} M_{nj}^2 = O(B_n^{-1}). \quad \square \tag{3.44}$$

The following central limit theorem under dependence, with conditions given in a form adapted to the present applications, is proved as Theorem 25.12 of SLT.

3.10 Theorem $\sum_{i=1}^{n} c_{ni} u_i \to_d \mathrm{N}(0, V)$ as $n \to \infty$ if

(a) $\mathrm{E}\left(\sum_{i=1}^{n} c_{ni} u_i\right)^2 \to V < \infty$
(b) $\{u_i\}_{i=1}^{n}$ satisfies Assumption **1.2**
(c) $\{c_{ni}\}_{i=1}^{n}$ satisfies Assumption **3.9**. $\square$

Comparing the conditions of Theorem **3.10** with those of SLT Theorem 25.12 itself, condition (a) of the latter theorem specifies a limiting L_2-norm of 1, but this is an arbitrary choice of normalization and any finite positive value can be substituted. Conditions (b) and (c) impose the conditions of SLT Theorem 25.6, which is a CLT for mixingales due to Robert de Jong (Theorem 2 of [22]).

Letting $\{\mathcal{F}_{ni}\}$ denote a filtration defined on the relevant probability space, an L_2-mixingale of size $-\varphi_0$ is a pair $\{x_{ni}, \mathcal{F}_{ni}\}$ satisfying the conditions

$$\begin{aligned} \|\mathrm{E}(x_{ni}|\mathcal{F}_{n,i-m})\|_2 &\le |c_{ni}|\zeta_m \\ \|x_{ni} - \mathrm{E}(x_{ni}|\mathcal{F}_{n,i+m})\|_2 &\le |c_{ni}|\zeta_{m+1} \end{aligned} \tag{3.45}$$

where $\zeta_m = O(m^{-\varphi})$ for $\varphi > \varphi_0$ and c_{ni} is an array of scale constants. The filtration is typically of the form $\mathcal{F}_{ni} = \sigma(x_{nj}, j \le i)$. The key fact is that the property specified in Assumption **1.2**(b), of L_2-near-epoch dependence (NED) on a mixing process, is sufficient for u_i to have the L_2-mixingale property of size $-\frac{1}{2}$.[2] With suitable scale adjustments, the property also extends to $c_{ni}u_i$ and the series under present consideration have the general form of $x_{ni} = c_{ni}u_i$ where $c_{ni} = a_{ni}(s,t)/\kappa(n)$.

A special feature of SLT Theorem 25.6 is that it accommodates heterogeneity in the distribution of the process increments by fine-tuning the specification of the blocking parameters. This is the role of Assumption **3.9**. Like many limit results for dependent processes, Theorem 25.6 of SLT works by breaking the sample series into r_n blocks of B_n successive terms, similarly to Theorem **2.10**. The property being exploited is that partial sums of mixingales behave approximately like martingales, with the residual terms becoming negligible as n increases under suitable restrictions on the range of dependence. This allows a central limit theorem for martingales[3] to be invoked, although with heterogeneity in the underlying series the blocking construction must satisfy Assumption **3.9**. This condition is critical in the antipersistent case, noting how the scale constants depend on d.

The CLT can then be shown, subject to the further condition that the normalized block sums form a uniformly square-integrable array. The condition on the blocks is essentially equivalent to what was previously shown for the increments of the form (3.18), with Assumption **3.1** needed to validate the step. Here too, antipersistent processes are subject to an additional restriction relative to long memory processes.[4]

[2] Proved as SLT Theorem 18.7.
[3] For example, SLT Theorem 25.3.
[4] Lemma 25.10 of SLT offers some additional insight into the role of these restrictions.

With these preliminaries established, a series of lemmas can be proved for the dependent shock case, to replace those given in §3.1–§3.4. The first, relating to $R_{3n}(s,t)$ defined in (3.2), is the generalization of Lemma **3.3**.

3.11 Lemma Under Assumption **1.2**, $\lim_{n\to\infty} \mathrm{E}(R_{3n}(s,t)^2) = O(N^{2d-1})$ as $N \to \infty$.

Proof By the arguments of Theorem **2.10** and Corollary **2.11**,

$$\begin{aligned}\mathrm{E}(R_{3n}(s,t))^2 &\sim \frac{\omega_u^2}{\kappa(n)^2}\sum_{i=-\infty}^{-nN} a_{ni}(s,t)^2 \\ &\sim \omega_u^2(s-t)^{2d+1}(\Upsilon_d - \Upsilon_d^N) \end{aligned} \tag{3.46}$$

where the order of magnitude in N of $\Upsilon_d - \Upsilon_d^N$ is given in (2.21). The equality in (3.4) is now replaced by the first limit result in (3.46) but the second step of the argument follows (3.4) directly. ∎

The next step is to prove a replacement for Lemma **3.4**, but to do this one further technical property has to be shown, which is that increments of the empirical process $R_{1n}(s,t) + R_{2n}(s,t)$ are asymptotically uncorrelated. This proof can conveniently be given first.

3.12 Theorem Let $z_{ni} = c_{ni}u_i$ where $c_{ni} = O(n^{-1/2})$ and for $r \in [-N,1]$ define $Z_n(r) = \sum_{i=-nN}^{[nr]} z_{ni}$ so that for $q > r$, $Z_n(q) - Z_n(r) = \sum_{i=[nr]+1}^{[nq]} z_{ni}$. If Assumption **1.2** holds, then for $-N \le r_1 < q_1 \le r_2 < q_2 \le 1$,

$$\lim_{n\to\infty} \mathrm{E}\left((Z_n(q_1) - Z_n(r_1))(Z_n(q_2) - Z_n(r_2))\right) = 0.$$

Proof For any $\eta \in (0, q_2 - r_2)$,

$$\begin{aligned}&|\mathrm{E}(Z_n(q_1) - Z_n(r_1))(Z_n(q_2) - Z_n(r_2))| \\ &\le \left|\sum_{i=[nr_1]+1}^{[nq_1]}\sum_{k=[nr_2]+1}^{[n(r_2+\eta)]} \mathrm{E}(z_{ni}z_{nk})\right| + \left|\sum_{i=[nr_1]+1}^{[nq_1]}\sum_{k=[n(r_2+\eta)]+1}^{[nq_2]} \mathrm{E}(z_{ni}z_{nk})\right|. \end{aligned} \tag{3.47}$$

Applying the modulus inequality to the first majorant term of (3.47) followed by the Cauchy-Schwarz inequality and then using the fact that $z_{nk} = O_p(n^{-1/2})$ and the assumption of weak dependence,

$$\left|\sum_{i=[nr_1]+1}^{[nq_1]}\sum_{k=[nr_2]+1}^{[n(r_2+\eta)]}\mathrm{E}(z_{ni}z_{nk})\right| \le \mathrm{E}\left|\sum_{i=[nr_1]+1}^{[nq_1]} z_{ni}\sum_{k=[nr_2]+1}^{[n(r_2+\eta)]} z_{nk}\right|$$
$$\le \left\|\sum_{i=[nr_1]+1}^{[nq_1]} z_{ni}\right\|_2 \left\|\sum_{k=[nr_2]+1}^{[n(r_2+\eta)]} z_{nk}\right\|_2 = O(\eta^{1/2}).$$

Next, consider the second majorant term of (3.47) where $|i-k| \ge n(r_2+\eta-q_1) \ge n\eta$. Assumption **1.2**(a) implies

$$\mathrm{E}(z_{ni}z_{nk}) = \gamma_u(|i-k|)c_{ni}c_{nk} = O\left((n(r_2+\eta-q_1))^{-1-\delta}n^{-1}\right) \tag{3.48}$$

and hence the second majorant term of (3.47) is of $O((n\eta)^{-\delta})$. Now, choose $\eta = O(n^{-\pi})$ for $0 < \pi < 1$ so that $(n\eta)^{-\delta} = O(n^{-\delta(1-\pi)})$ and

$$|\mathrm{E}(Z_n(q_1)-Z_n(r_1))(Z_n(q_2)-Z_n(r_2))| = O(\max\{n^{-\pi/2}, n^{-\delta(1-\pi)}\}) = o(1). \qquad \blacksquare$$

For the present application of Theorem **3.12** let $c_{ni} = a_{ni}(s,t)/\kappa(n)$, which satisfies the specified order of magnitude according to Theorem **2.5** and (2.27), so that $\sum_{i=[nr]+1}^{[nq]} z_{ni}$ is an increment of the process $R_{1n}(s,t) + R_{2n}(s,t)$ in (3.2). When $d < 0$ there are points i detailed in the proof of Lemma **3.4** such that $c_{ni} = O(n^{-1/2-d})$, contradicting the assumption of the theorem. However, these contributions are of small order in sums of $O(n)$ terms and omitting them does not affect the distribution of the aggregate.

3.13 Lemma If $|d| < \frac{1}{2}$ and Assumption **1.2** holds,

$$R_{1n}(s,t) + R_{2n}(s,t) \xrightarrow{\mathrm{d}} \mathrm{N}\left(0, \omega_u^2\Upsilon_d^N(s-t)^{2d+1}\right)$$

for any s and t with $0 \le t < s \le 1$.

Proof As before, the case shown is $s = 1$ and $t = 0$. Appealing to Theorem **3.10**, the main task is to verify that Assumption **3.9** holds for the scale constants when $|d| < \frac{1}{2}$. In $R_{1n}(1,0)$, the slowly varying components cancelling in the limit,

$$|c_{ni}| = \frac{|a_{ni}(1,0)|}{\kappa(n)} \sim \frac{(n-i)^d}{n^{d+1/2}} \tag{3.49}$$

with 0^d assigned the value 1. When $d > 0$, c_{ni} is decreasing in i and according to (3.42),

$$M_{nj} \sim \frac{(n-(j-1)B_n)^d}{n^{d+1/2}} \sim \frac{(r_n - j + 1)^d}{B_n^{1/2} r_n^{d+1/2}}.$$

Hence,

$$\max_{1 \le j \le r_n+1} M_{nj} = M_{n1} = O(n^{-1/2}) = o(B_n^{-1/2}) \tag{3.50}$$

and

$$\sum_{j=1}^{r_n} M_{nj}^2 \backsimeq \frac{1}{B_n r_n^{2d+1}} \sum_{j=1}^{r_n} (r_n - j + 1)^{2d} = O(B_n^{-1}). \tag{3.51}$$

These results hold for $B_n = [n^{1-\alpha}]$ for any $\alpha \in (0,1]$. If $d < 0$ then c_{ni} is increasing in i and takes its maximum at $i = n$, at which point $a_{ni} = 1$. In this case let $B_n = [n^{1-\alpha}]$ for $-2d < \alpha < 1$, which is a feasible choice with $d > -\frac{1}{2}$. Then,

$$\max_{1 \le j \le r_n+1} M_{nj} = c_{nn} \sim \frac{1}{n^{d+1/2}} = o(B_n^{-1/2}). \tag{3.52}$$

In (3.51), j replaces $j-1$ in the definition of M_{nj} but the sum is still of $O(B^{-1})$. The conditions of Theorem **3.10** are accordingly satisfied for $R_{1n}(1,0)$ with $|d| < \frac{1}{2}$. The limiting variance V in Theorem **3.10**(a) is defined by (3.5) with ω_u^2 replacing σ_u^2 by application of Theorem **2.10**.

Next, consider the sum $R_{2n}(1,0)$ in which the terms are labelled $1 - nN \le i \le 0$. Similarly to (3.15), break this sum into N blocks each of length n, denoted

$$R_{2n}^k(1,0) = \sum_{i=1-n(k+1)}^{-nk} \frac{a_{ni}(1,0)}{\kappa(n)} u_i, \quad k = 0, \dots, N-1.$$

Assumption **3.9** must be verified for the case c_{ni} for $i = 1 - n(k+1), \dots, -nk$ and according to (2.31), in this instance,

$$|c_{ni}| \sim \frac{|(n-i)^d - (-i)^d|}{n^{d+1/2}}. \tag{3.53}$$

To distinguish the different cases write M_{nj}^k for $k = 0, \ldots, N-1$, where

$$\begin{aligned} M_{nj}^k &= \max_{1-nk-jB_n \le i \le -nk-(j-1)B_n} |c_{ni}| \\ &\sim \frac{|(n+nk+(j-1)B_n)^d - (nk+(j-1)B_n)^d|}{n^{d+1/2}}. \end{aligned} \tag{3.54}$$

For both $d > 0$ and $d < 0$, c_{ni} is decreasing as $-i$ increases and M_{nj}^k is maximized over $j = 1, \ldots, r_n$ at $j = 1$.

Consider first the cases $k > 0$. For each of these,

$$\max_{1 \le j \le r_n+1} M_{nj}^k \sim \frac{n^d|(k+1)^d - k^d|}{n^{d+1/2}} = O(n^{-1/2}) = o(B_n^{-1/2}). \tag{3.55}$$

Making the change of variable $\tau = k+(j-1)/r_n$, it is also found from (3.54) that

$$\begin{aligned} \sum_{j=1}^{r_n} (M_{nj}^k)^2 &\sim \frac{B_n^{2d}}{n^{2d+1}} \sum_{j=1}^{r_n} \left((r_n(k+1)+j-1)^d - (r_n k + j - 1)^d \right)^2 \\ &\sim \frac{1}{B_n} \int_k^{k+1} \left((1+\tau)^d - \tau^d \right)^2 \mathrm{d}\tau = O(B_n^{-1}). \end{aligned} \tag{3.56}$$

It follows that for each $k > 0$ the conditions of Theorem **3.10** are satisfied by $R_{2n}^k(1,0)$ and applying Theorem **2.10**,

$$V = \lim_{n\to\infty} \mathrm{E}(R_{2n}^k(1,0))^2 = \omega_u^2 \int_k^{k+1} \left((1+\tau)^d - \tau^d \right)^2 \mathrm{d}\tau. \tag{3.57}$$

When $d > 0$, the indicated orders of magnitude in (3.55) and (3.56) also hold for $k = 0$. However, when $d < 0$, $k^d > (k+1)^d$ and in the case $k = 0$, (3.55) has to be replaced by

$$\max_{1 \le j \le r_n+1} M_{nj}^0 = M_{n1}^0 \sim \frac{1}{n^{d+1/2}} = O(n^{-d-1/2}) = o(B_n^{-1/2})$$

where the last equality holds provided that $B_n = [n^{1-\alpha}]$ for $-2d < \alpha < 1$. The calculation in (3.56) continues to apply in this case for $k = 0$, so that (3.44) is still validated.

Thus, the individual blocks R_{1n} and $R_{2n}^0, \ldots, R_{2n}^{N-1}$ are shown to be Gaussian in the limit. They are also uncorrelated in the limit in view of Theorem **3.12**.

However, this means that the product of the limiting marginal densities specifies their joint distribution. The blocks are accordingly independent in the limit, so that in particular their sum is Gaussian. The variances of the form (3.57) for each k are summable over $k = 0, \ldots, N-1$ as $N \to \infty$, hence the sum approaches the limit having the form of (3.7) with ω_u^2 replacing σ_u^2. The extension of these arguments to the cases of $R_{1n}(s,t)$ and $R_{2n}(s,t)$, for any $s,t \in (0,1]$ follows the scheme set out for Lemma **3.4**. ∎

The property of independent increments is important for another reason, which is to show that the dependence of the limit process X is determined exclusively by d, in the manner represented by equation (2.1), so validating Corollary **2.8** and the characteristic fractional dependence illustrated by (2.17). Such a demonstration was not needed in Lemma **3.4** where the increments were independent by assumption.

The next step is to show uniform boundedness in probability as the preliminary to proving uniform tightness. Theorem **3.5** is for independent shocks and does not apply under Assumption **1.2**. However, the following supplement to Assumption **1.2** is shown to be a sufficient restriction on the dependence.

3.14 Assumption For $t \in [-N, 1]$, $\delta \in (0, 1-t]$, $n \in \mathbb{N}$, and $\eta > 0$,

$$P\left(\sup_{t \le s \le t+\delta} |u_{[ns]}| \le \eta\right) \ge (1 - P(|u_i| > \eta))^{[n(t+\delta)]-[nt]+1}. \quad \square \tag{3.58}$$

If the second equality of (3.20) in the proof of Theorem **3.5** is replaced by inequality (3.58) with $\eta = \kappa(n)\varepsilon$, the sufficiency part of Theorem **3.5** continues to hold under dependence. This is shown formally as follows.

3.15 Corollary Under Assumptions **1.2**, **3.1**, and **3.14**, for $\delta > 0$ and $0 \le t \le 1-\delta$ the collection $\{\widetilde{T}_n^2(t,\delta), n \in \mathbb{N}\}$ is uniformly bounded in probability, where $\widetilde{T}_n(t,\delta)$ is defined in (3.18).

Proof The proof of Theorem **3.5** can be followed at most points, but under Assumption **3.14**, the second equality in (3.20) is replaced by an inequality as in (3.58). Equation (3.23) is therefore replaced by

$$-\log P\left(\sup_{-N \le s \le 1} \frac{|u_{[ns]}|}{\kappa(n)} \le \varepsilon\right) \le O(\varepsilon^{-r} \log(\varepsilon\kappa(n))^{-1-\mu}). \tag{3.59}$$

The probability on the left-hand side of (3.59) must converge to 1 as $\varepsilon \to \infty$. ∎

If the inequality in (3.59) was not imposed the divergence of the supremum with positive probability could occur, but Assumption **3.14** rules out this eventuality.

Having regard to the plausibility of condition (3.58), the first thing to observe is that the existence or otherwise of autocorrelation in levels is irrelevant. The relation concerns only the joint distribution of the absolute values, or squares, of the process. Consider partitioning the sequence elements u_i in (3.58) into subsets S_A and S_B and so defining events $A = \bigcup_{S_A}\{|u_i| > \eta\}$ and $B = \bigcup_{S_B}\{|u_i| > \eta\}$. The intersection of the complements of these two sets is the event whose probability is on the left-hand side of (3.58). The relation

$$P(A^c \cap B^c) \geq P(A^c)P(B^c) = (1 - P(A))(1 - P(B)) \qquad (3.60)$$

rearranges using the de Morgan law as $P(B|A) \geq P(B)$ and equivalently as $P(A|B) \geq P(A)$. If an element of one set exceeds the bound, it is at least as probable according to (3.60) that an element of the other set will do so. Consider circumstances in which this inequality might be reversed. Suppose for example that the elements of S_A are odd numbered in the time sequence and those of S_B are even numbered. In this case a failure of (3.60) would represent the opposite of volatility clustering. Negative correlation of adjacent squares in the time sequence is not an impossible scenario, but few empirical studies of conditional heteroscedasticity suggest such outcomes.

Further, the partitioning exercise can be performed on each of sets S_A and S_B and a further inequality obtained for the minorant of (3.60). Repeating the cycle until all the sets are singletons defines a chain of inequalities for products of probabilities, terminating in the right-hand side of (3.58). If (3.60) held for every such partition then Assumption **3.14** certainly follows, but this strong condition is only one of many ways the assumption might be validated. In the example of shocks adjacent in time, note how the succeeding partitions must involve sample elements further separated in time and hence with a diminishing connection under weak dependence. For Assumption **3.14** to be contradicted, inequality (3.60) would need to fail at a preponderance of the partitioning steps. These considerations suggest that the alternative to (3.58) can with some confidence be relegated to a special case.

A further issue is that Lemma **3.6** cannot be applied directly to show the condition imposed by Theorem **3.8**, since it uses Theorem **A.8** which does not work for mixingales. A uniform integrability proof for mixingales due to Donald McLeish ([52], [53]) is given as Theorem 17.14 of SLT. The drawback with this theorem is that while it shows uniform integrability of the maximal sum it does not establish a rate of convergence in the manner of equation (3.25), which does need to be known to fulfil assumption (b) of Theorem **3.8**.

While the proof given for Theorem 17.14 of SLT is both lengthy and tricky, the extension is minor and can be given in terms of steps in that proof. Similarly to Theorem **3.10**, the details can be pursued with the relevant pages of SLT to hand (pages 357–360 of the 2nd edition). Except for the addition of condition (3.61), the following statement matches SLT Corollary 17.15, which is the extension to arrays of the main result for mixingale sequences. In the present application, x_{ni}/c_{ni} corresponds to u_i, which is stationary and L_r-bounded for $r \geq 2$ under Assumption **1.2**.

3.16 Theorem For $r \geq 2$ let $\{x_{ni}, \mathcal{F}_{ni}\}$ be a L_r -bounded, L_2-mixingale array of size $-\frac{1}{2}$. Let $S_{nk} = \sum_{i=1}^{k} x_{ni}$ for $1 \leq k \leq n$ and $v_n^2 = \sum_{i=1}^{n} c_{ni}^2$ where the c_{ni} are the mixingale scale constants. If the array $\{x_{ni}^2/c_{ni}^2\}$ is uniformly integrable the collection $\{\max_{1\leq k\leq n} S_{nk}^2/v_n^2, n \in \mathbb{N}\}$ is uniformly integrable, with

$$\mathrm{E}\left(\left(\max_{1\leq k\leq n} S_{nk}^2/v_n^2\right) 1_{\{\max_{1\leq k\leq n}|S_{nk}|/v_n>\eta\}}\right) = o(\eta^{2-r}). \tag{3.61}$$

Proof In view of SLT's Theorem 17.14, it suffices to show that equation (3.61) also holds under the same conditions. The theorem's proof of uniform integrability proceeds by setting a number ε to be the common upper bound of three terms, there denoted $\mathrm{E}(\hat{u}_n^2)$, $\mathrm{E}(\hat{z}_n^2)$, and $\mathcal{E}_{M/6}(\hat{y}_n^2)$. The sum of these terms defines a bound on $\mathcal{E}_M(\hat{x}_n^2) = \mathrm{E}(\hat{x}_n^2 1_{\{|\hat{x}_n|>M\}})$ where $\hat{x}_n^2 = \max_{1\leq k\leq n} S_{nk}^2/v_n^2$. It is shown that ε is a decreasing function of the bound M with $\varepsilon \to 0$ as $M \to \infty$.

The key relations are (17.65), (17.67), and (17.74). Equation (17.65) shows $\mathrm{E}(\hat{u}_n^2) = O(m^{-\delta})$, where m is the order of lead/lag as in (3.45), defining the mixingale truncation in equations (17.52)–(17.54). The parameter δ can be equated with the serial dependence parameter from Assumption **1.2**(a). The notation $\backsimeq$ is used here to denote matching orders of magnitude as $\varepsilon \to 0$. Following (17.65), let the bound be represented as $\varepsilon \backsimeq m^{-\delta}$ and equivalently, let $m \backsimeq \varepsilon^{-1/\delta}$ denote the approximate value of m required to attain the bound.

Next, under the assumption that x_{ni} is an L_r-bounded random variable for $r \geq 2$, use the result $\mathrm{E}\left((x_{ni}/c_{ni})^2 1_{\{|X_{ni}/c_{ni}|>B\}}\right) = o(B^{2-r})$ from Theorem **A.5** of Appendix A. With this assumption, inequality (17.67) (where X_t/c_t can be read as x_{ni}/c_{ni}) can be extended by a step beyond the source proof and written as

$$\mathrm{E}(\hat{z}_n^2) \ll mB^{2-r} \backsimeq \varepsilon. \tag{3.62}$$

Since the relations represent orders of magnitude the notation allows scale constants to be taken as implicit.

Substituting for m in (3.62) and rearranging gives $B \backsimeq \varepsilon^{(1+1/\delta)/(2-r)}$ as approximating the value of B needed to attain the bound. Inequality (17.74) can be rendered in the same manner as

$$\mathcal{E}_{M/6}(\hat{y}_n^2) \ll \frac{m^4 B^4}{M} \backsimeq \varepsilon. \tag{3.63}$$

Substituting for m and B in (3.63), rearranging, simplifying, and inverting gives for $0 < \varepsilon < 1$ and $r \geq 2$,

$$M^{2-r} \backsimeq \varepsilon^{2+r+4(r-1)/\delta} \leq \varepsilon. \tag{3.64}$$

Equating M with η to match the notation of Theorem **A.8**, Theorem **A.5** applied to the random variable $\max_{1\leq k\leq n} |S_{nk}|^2/v_n^2$ leads to (3.61) replacing (17.75). ∎

Similarly to Theorem **A.8**, the conditions of Theorem **3.16** do not restrict the ordering of the weights c_{ni}. As is apparent from (3.45), the dependence characteristics of the normalized elements x_{ni}/c_{ni}, which correspond to the stationary sequence u_i in the present application, do not depend on the c_{ni}. Provided their sum of squares does not change, these can be permuted arbitrarily in the manner described in the proof of Lemma **3.6**.

The required result to replace Lemma **3.6** is as follows, with $\widetilde{T}_n(t, \delta)$ defined as in (3.18) although taking note that ω_u^2 replaces σ_u^2 in the variance formulae following (3.16).

3.17 Lemma Under Assumptions **1.2**, **3.1**, and **3.14**, the collection $\{\widetilde{T}_n^2(t, \delta), n > n_0\}$ is uniformly integrable and satisfies (3.25), for $n_0 < \infty$ and all $0 < \delta < 1$ and $t \in [0, 1 - \delta]$.

Proof This is identical to the proof of Lemma **3.6**, subject only to replacement of the citations of Theorem **A.8**, including that in Corollary **A.9**, by citations of Theorem **3.16**. The array elements x_{ni} in the notation of the latter theorem correspond to $c_{np(i)}u_i$ where $c_{np(i)}$ denotes any of the permutations defined in the proof of Lemma **3.6**. In the application, the array $\{x_{ni}^2/c_{ni}^2\}$ specified in Theorem **3.16** corresponds to the set $\{u_i^2\}$ and according to Corollary **3.15**, its uniform integrability follows from Theorem **A.5** and the assumptions. ∎

Gathering up the necessary additional lemmas, the main result of this section can now be given for X_n in (2.28) where, as before, X is the fBm in (2.1).

3.18 Theorem If $|d| < \frac{1}{2}$ and Assumptions **1.2**, **3.1**, and **3.14** hold then $X_n \to_d X$.

Proof Lemma **3.11** focuses attention as before on the terms R_{1n} and R_{2n} of (3.2). Then, Theorem **3.12** and Lemma **3.13** establish the Gaussianity and increment independence of the limit distribution. Lemma **3.17** establishes assumption (b) of Theorem **3.8**, which proves the a.s. continuity of the limit distribution in the same way as before. Theorem **2.10** and Corollary **2.11** fix the covariance structure to complete the proof. ∎

The conclusion of these results is that the one consequence for the limit distribution of swapping Assumption **1.1** for Assumptions **1.2** and **3.14** is the replacement of the variance parameter. However, as can be deduced from the style of the proofs, the rate of convergence to the limit can be substantially slower. Much the most important consequence of shock dependence is that larger samples may be needed to attain comparable approximations to the limit.

3.6 The Multivariate FCLT

The matrix notation required to handle the multivariate case is detailed in §2.5. Let D^m denote the space of m-dimensional càdlàg functions and C^m the space of m-dimensional continuous functions. For clarity of notation the domain of the functions is not indicated explicitly here, although this is $[0, 1]$ in the present case.

For the space of m-tuples endowed with the Skorokhod topology (see page 23 for details), the notation D^m is taken to imply that the times of discontinuities are coordinated. In other words, Skorokhod distances must be defined in terms of a common change-of-time function λ. Vectors of càdlàg functions whose jumps are uncoordinated are not treated in the present theory. However, the elements $X_{1n}, \ldots, X_{mn}$ in the present case are step functions whose jump times are dictated by observation dates so that coordination is not an issue in general.

The following fundamental result, proved as Theorem 30.13 of SLT, extends the well-known Cramér-Wold theorem to random functions and is at the heart of the multivariate FCLT, so deserves a formal statement in this context.

3.19 Theorem Let $\boldsymbol{X}_n \in D^m$ be an m-vector of random elements. $\boldsymbol{X}_n \to_d \boldsymbol{X}$ where $P(\boldsymbol{X} \in C^m) = 1$ if and only if $\boldsymbol{\lambda}'\boldsymbol{X}_n \to_d \boldsymbol{\lambda}'\boldsymbol{X}$, where $P(\boldsymbol{\lambda}'\boldsymbol{X} \in C) = 1$, for every fixed $\boldsymbol{\lambda}$ ($m \times 1$) with $\boldsymbol{\lambda}'\boldsymbol{\lambda} = 1$. □

With this setup the multivariate extension of Theorem **3.18** can be specified. Taking advantage of the developments in §3.5, it is possible to allow the conditions of Assumption **1.2** for full generality, subject to the following.

3.20 Assumption

(a) In $\boldsymbol{D}_n$ defined in (2.46), $-\frac{1}{2} < d_1 \leq \cdots \leq d_m < \frac{1}{2}$.

(b) The elements of $\{\boldsymbol{u}_i\}$ individually satisfy Assumptions **1.2**, **3.1**, and **3.14** and have covariance matrix $\boldsymbol{\Omega}_u$ as defined in (2.74) that is finite and positive definite. □

3.21 Theorem If the process $\boldsymbol{X}_n$ ($m \times 1$) is defined by (2.48) and Assumption **3.20** holds, then $\boldsymbol{X}_n \to_d \boldsymbol{X}$ where $\boldsymbol{X}$ is a vector whose k^{th} element is a fBm with parameter d_k, with covariance structure given by the limit in (2.75).

Proof By Theorem **3.19** the result follows if $\boldsymbol{\lambda}'\boldsymbol{X}_n$ converges in distribution to an a.s. continuous Gaussian limit $\boldsymbol{\lambda}'\boldsymbol{X}$ for all m-vectors $\boldsymbol{\lambda}$ of unit length. These limits are not fBms in general, but as a consequence of Corollary **2.12** the cases $\boldsymbol{\lambda} = \boldsymbol{e}_k$ for $k = 1, \ldots, m$ (the columns of the identity matrix of order m) yield fBms with parameters d_k.

Establishing the finite-dimensional distributions can proceed in tandem with the arguments of Theorem **3.18**. In parallel with (3.2) let

$$\begin{aligned}\boldsymbol{\lambda}'\boldsymbol{X}_n(s) - \boldsymbol{\lambda}'\boldsymbol{X}_n(t) &= \left(\sum_{i=[nt]+1}^{[ns]} + \sum_{i=1-nN}^{[nt]} + \sum_{i=-\infty}^{-nN} \right) \boldsymbol{\lambda}'\boldsymbol{D}_n^{-1}\boldsymbol{A}_{ni}(s,t)\boldsymbol{u}_i \\ &= R_{1n}^{\lambda}(s,t) + R_{2n}^{\lambda}(s,t) + R_{3n}^{\lambda}(s,t). \end{aligned} \tag{3.65}$$

To verify the status of the remainder term $R^{\lambda}_{3n}(s,t)$, note that according to (2.50), $\Upsilon_{kl} = \Upsilon^N_{kl} + (\Upsilon_{kl} - \Upsilon^N_{kl})$ where

$$\begin{aligned}\Upsilon_{kl} - \Upsilon^N_{kl} &= \int_N^{\infty} ((1+\tau)^{d_k} - \tau^{d_k})((1+\tau)^{d_l} - \tau^{d_l})\mathrm{d}\tau \\ &\leq \max_{k,l}\left\{\int_N^{\infty} ((1+\tau)^{d_k} - \tau^{d_k})^2\mathrm{d}\tau, \int_N^{\infty} ((1+\tau)^{d_l} - \tau^{d_l})^2\mathrm{d}\tau\right\} \\ &= O(N^{2\max\{d_k,d_l\}-1}).\end{aligned}$$

It follows by Theorem **2.9** that

$$\begin{aligned}\lim_{n\to\infty} \mathrm{E}(R^{\lambda}_{3n}(s,t)^2) &= \sum_k\sum_l \lambda_k\lambda_l\omega_{kl}(\Upsilon_{kl} - \Upsilon^N_{kl})(s-t)^{d_k+d_l+1} \\ &= O(N^{2d_m-1}).\end{aligned}$$

As in the univariate case, N can be chosen large enough under Assumption **3.20**(a) for R^{λ}_{3n} to be negligible in L_2 -norm.

Consider the remaining sums in turn. As before, it is convenient to consider the exemplar case $s = 1$ and $t = 0$. As in Lemma **3.13**, the finite-dimensional CLT is proved by verifying the conditions of Theorem **3.10**, the essential step being to specify the scale constants to be attached to each observation. For fixed $\boldsymbol{\lambda}$, write $W^{\lambda}_{ni} = \boldsymbol{\lambda}'\boldsymbol{D}_n^{-1}\boldsymbol{A}_{ni}(1,0)\boldsymbol{u}_i$ and so define the array $\{c^{\lambda}_{ni}\}$, being the positive square root of

$$(c^{\lambda}_{ni})^2 = \mathrm{E}((W^{\lambda}_{ni})^2) = \boldsymbol{\lambda}'\boldsymbol{A}_{ni}(1,0)\boldsymbol{D}_n^{-1}\boldsymbol{\Omega}_u\boldsymbol{D}_n^{-1}\boldsymbol{A}_{ni}(1,0)\boldsymbol{\lambda}. \tag{3.66}$$

Then, for $i = 1,\dots,n$ corresponding to the terms of $R^{\lambda}_{1n}(1,0)$, applying Theorem **2.5** and noting that the slowly varying factors cancel in the limit,

$$(c^{\lambda}_{ni})^2 \sim \sum_{k=1}^{m}\sum_{l=1}^{m} \lambda_k\lambda_l\omega_{kl}\frac{(n-i)^{d_k+d_l}}{n^{d_k+d_l+1}} = \frac{(n-i)^{2d_1}}{n^{2d_1+1}}J^{\lambda}_{ni} \tag{3.67}$$

with $d_1 = \min\{d_1,\dots,d_m\}$ and

$$J^{\lambda}_{ni} = \sum_{k=1}^{m}\sum_{l=1}^{m} \lambda_k\lambda_l\omega_{kl}\left(\frac{n-i}{n}\right)^{d_k+d_l-2d_1}. \tag{3.68}$$

One difference between c^{λ}_{ni} and the formula in (3.49) is that here the shock covariance matrix elements need to be incorporated into the scale constants, whereas in the scalar context the variance can conveniently be separated from the scale functions. The exponents in (3.68) are nonnegative by construction and Assumption **3.20**(b) ensures the array J^{λ}_{ni} is positive and finite for all $n > 1$. The conditions of Assumption **3.9** are therefore met, following the line of reasoning in the proof of Lemma **3.13**. The finite-dimensional distributions of the processes R^{λ}_{1n} are proved in the same manner as before.

For $i = 1 - nN, \ldots, 0$, corresponding to terms of $R^{\lambda}_{2n}(1, 0)$,

$$
\begin{aligned}
(c^{\lambda}_{ni})^2 &\sim \sum_{k=1}^{m}\sum_{l=1}^{m} \lambda_k\lambda_l\omega_{kl}\frac{((n-i)^{d_k} - (-i)^{d_k})((n-i)^{d_l} - (-i)^{d_l})}{n^{d_k+d_l+1}} \\
&= \frac{(n-i)^{2d_1}}{n^{2d_1+1}}J^{\lambda 1}_{ni} + \frac{(-i)^{2d_1}}{n^{2d_1+1}}J^{\lambda 2}_{ni} - 2\frac{(n-i)^{d_1}(-i)^{d_1}}{n^{2d_1+1}}J^{\lambda 3}_{ni}
\end{aligned} \tag{3.69}
$$

where

$$
\begin{aligned}
J^{\lambda 1}_{ni} &= \sum_{k=1}^{m}\sum_{l=1}^{m} \lambda_k\lambda_l\omega_{kl}\left(\frac{n-i}{n}\right)^{d_k+d_l-2d_1} \\
J^{\lambda 2}_{ni} &= \sum_{k=1}^{m}\sum_{l=1}^{m} \lambda_k\lambda_l\omega_{kl}\left(\frac{-i}{n}\right)^{d_k+d_l-2d_1} \\
J^{\lambda 3}_{ni} &= \frac{1}{2}\sum_{k=1}^{m}\sum_{l=1}^{m} \lambda_k\lambda_l\omega_{kl}\left(\left(\frac{n-i}{n}\right)^{d_k-d_1}\left(\frac{-i}{n}\right)^{d_l-d_1} + \left(\frac{-i}{n}\right)^{d_k-d_1}\left(\frac{n-i}{n}\right)^{d_l-d_1}\right).
\end{aligned}
$$

By assumption the terms $J^{\lambda 1}_{ni}$, $J^{\lambda 2}_{ni}$, and $J^{\lambda 3}_{ni}$ are positive and finite for all $1 - nN \le i \le 0$ and all $n \ge 1$. There therefore exist constants L^{λ} and U^{λ}, such that

$$
0 < L^{\lambda} \le \min(J^{\lambda 1}_{ni}, J^{\lambda 2}_{ni}, J^{\lambda 3}_{ni}) \le \max(J^{\lambda 1}_{ni}, J^{\lambda 2}_{ni}, J^{\lambda 3}_{ni}) \le U^{\lambda} < \infty
$$

and such that when n is large enough, $L^{\lambda}_{ni} \le (c^{\lambda}_{ni})^2 \le U^{\lambda}_{ni}$ where

$$
\begin{aligned}
L^{\lambda}_{ni} &= L^{\lambda}\frac{((n-i)^{d_1} - (-i)^{d_1})^2}{n^{2d_1+1}}, \\
U^{\lambda}_{ni} &= U^{\lambda}\frac{((n-i)^{d_1} - (-i)^{d_1})^2}{n^{2d_1+1}}.
\end{aligned}
$$

Consider the stationary weakly dependent process $W^{\lambda}_{ni}/c^{\lambda}_{ni}$, for which scale constants of unity are appropriate. The argument of Lemma **3.13** applies to the increment processes $\sqrt{L^{\lambda}_{ni}}W^{\lambda}_{ni}/c^{\lambda}_{ni}$ and $\sqrt{U^{\lambda}_{ni}}W^{\lambda}_{ni}/c^{\lambda}_{ni}$, setting $\sqrt{L^{\lambda}_{ni}}$ and $\sqrt{U^{\lambda}_{ni}}$ respectively for the arrays of scale constants. In particular, the fact that the conditions of Assumption **3.9** are satisfied in these cases for $d_1 > -\frac{1}{2}$ was established in the proof of Lemma **3.13**, leading to equations (3.55) and (3.56). Since $\sqrt{L^\lambda_{ni}/c^\lambda_{ni}} < 1 < \sqrt{U^\lambda_{ni}/c^\lambda_{ni}}$, W^λ_{ni} itself is tending to a convex combination of these processes as n increases. If both $\sum_{i=1-N_n}^0 \sqrt{L^\lambda_{ni} W^\lambda_{ni}/c^\lambda_{ni}}$ and $\sum_{i=1-N_n}^0 \sqrt{U^\lambda_{ni} W^\lambda_{ni}/c^\lambda_{ni}}$ are Gaussian in the limit then the same is true of R^λ_{2n}. These arguments establish the finite-dimensional distributions of the scalar processes.

To prove uniform tightness of the distributions, it is convenient to define

$$T_{nk}(s,t) = \sum_{i=1-nN}^{[ns]} \frac{a_{nki}(s,t)}{n^{d_k+1/2}L_k(n)} u_{ki} \tag{3.70}$$

so that $\sum_{k=1}^m \lambda_k T_{nk}(s,t) = R^\lambda_{1n}(s,t) + R^\lambda_{2n}(s,t)$ from (3.65). Also let

$$\nu^\lambda_n(t,\delta)^2 = \sum_{k=1}^m \sum_{l=1}^m \lambda_k \lambda_l \omega_{kl} \sum_{i=1-nN}^{[n(t+\delta)]} \frac{a_{nki}(t+\delta,t)a_{nli}(t+\delta,t)}{n^{d_k+d_l+1}L_k(n)L_l(n)}. \tag{3.71}$$

Application of Corollary **2.12** gives

$$\lim_{n\to\infty} \nu^\lambda_n(t,\delta)^2 = \sum_{k=1}^m \sum_{l=1}^m \lambda_k \lambda_l \omega_{kl} \Upsilon_{kl} \delta^{d_l+d_k+1}. \tag{3.72}$$

It has to be shown that for $0 < \delta < 1$ and $t \in [0, 1-\delta]$ the collections

$$\left\{ \sup_{t\le s\le t+\delta} \frac{\left(\sum_{k=1}^m \lambda_k T_{nk}(s,t)\right)^2}{\nu^\lambda_n(t,\delta)^2}, n \ge n_0 \right\} \tag{3.73}$$

satisfy the conditions of Lemma **3.17**. If so, uniform tightness follows by Theorem **3.8** applied in the case $Y_n = \boldsymbol{\lambda}'\mathbf{X}_n$. Substituting $\delta^{d_l+d_k+1} = \delta^{(d_l+d_k-2d_1)}\delta^{2d_1+1}$ in (3.72) where the first factor has a nonnegative exponent, assumption (a) of Theorem **3.8** is satisfied if the members of the collection (3.73) play the role of $\widetilde{T}_n$ in (3.35).

To investigate (3.73), the approach is first to consider the terms of the sum one at a time. For $k = 1, \ldots, m$, the particular case of $(c^\lambda_{ni})^2$ in (3.66) when $\boldsymbol{\lambda} = \boldsymbol{e}_k$ (the kth column of the identity matrix) has the form

$$c_{nki}^2(s,t) = \frac{\omega_{kk} a_{nki}(s,t)^2}{n^{2d_k+1} L_k(n)^2}. \tag{3.74}$$

For each k, form the sums of these elements for the increment of width δ and note that

$$\nu_{nk}^2(t,\delta) = \sum_{i=1-nN}^{[n(t+\delta)]} c_{nki}^2(t+\delta, t) \to \omega_{kk} \Upsilon_{kk} \delta^{2d_k+1} \tag{3.75}$$

as $n \to \infty$. Under Assumptions **1.2**, **3.1**, and **3.14**, for each k the collections

$$\left\{ \sup_{t \le s \le t+\delta} \frac{T_{nk}^2(s,t)}{\nu_{nk}^2(t,\delta)}, n \ge n_0 \right\} \tag{3.76}$$

satisfy the conditions of Lemma **3.17**. For each of these, the conditions of Theorem **3.16** are satisfied under Assumption **1.2**, noting that the squared terms of the sum (3.70) divided by the $c_{nki}^2(s,t)$ from (3.74) have the form u_{ki}^2/ω_{kk}. These correspond to the x_{ni}^2/c_{ni}^2 in the notation of Theorem **3.16** and are uniformly integrable by Theorem **A.5**.

Given these properties for each of the m terms represented by (3.76), $m-1$ successive applications of Theorem **A.7** in conjunction with the argument leading to inequality (3.27) and followed by Theorem **A.6**, extend the required properties to the collection

$$\left\{ \sup_{t \le s \le t+\delta} \left(\sum_{k=1}^{m} \frac{\lambda_k T_{nk}(s,t)}{\lambda_k \nu_{nk}(t,\delta)} \right)^2, n \ge n_0 \right\} \tag{3.77}$$

where the factors λ_k have been harmlessly included in top and bottom of the ratios.

The next step is to apply this reasoning to another case, in which, for the given set of weights $\lambda_1, \ldots, \lambda_m$, u_{ki} in (3.70) is replaced for each $k = 1, \ldots, m$ by

$$u_{ki}^* = \frac{\lambda_k \nu_{nk}(t,\delta)}{\nu_n^\lambda(t,\delta)} u_{ki}. \tag{3.78}$$

Let $T_{nk}^*(s,t)$ denote this case of (3.70) with u_{ki}^* from (3.78) replacing u_{ki}. Then, (3.75) together with (3.72) shows that for any $\delta > 0$ the ratios of scale factors in (3.78) converge to finite constant limits. Hence, under Assumptions **1.2**, **3.1**, and **3.14** the collection

$$\left\{ \sup_{t \le s \le t+\delta} \left(\sum_{k=1}^{m} \frac{\lambda_k T^*_{nk}(s,t)}{\lambda_k v_{nk}(t,\delta)} \right)^2, n \ge n_0 \right\} \tag{3.79}$$

satisfies the required conditions in the same way as the collection (3.77).

This completes the proof, since the collection in (3.79) is identical with the collection in (3.73). ∎

Chapter 4
The Fractional Covariance

The material of the foregoing chapters lays the groundwork for posing various fundamental questions arising in econometrics, in particular, the distributions of regression coefficients when the variables entering the relationships in question may be fractional series, either regressors or disturbance terms, or both. Interest therefore focuses on the limit distributions of sample means, variances, and covariances under suitable normalizations.

Regressions may involve stationary series as defined by relations (1.1) and (1.2), but the cases of greatest interest inevitably involve partial sum processes of the form (2.22). This generalizes the usual cointegration methodology for nonstationary series with autoregressive roots of unity, which corresponds in the fractional context to the case $d = 0$. The suitably normalized covariance between a unit root process and a weakly dependent process is well known to converge to a stochastic integral of the Itô type, having Brownian motion as the integrator process.

The aim is now to generalize these results to the case where either or both of the partial sum and the stationary integrator are fractional. The development occupies the current chapter and Chapters 5 and 6 following, the three stages of the analysis being to calculate limiting expectations for the random variable in question, to develop heuristic representations of the limiting integrals, and then, not least, to prove weak convergence to these limits. The developments are based on some original ideas sketched in [20], which is joint work with Nigar Hashimzade. Chapter 7 then uses elementary bivariate models to illustrate what the results obtained imply for regression analysis.

4.1 Assumptions and Preliminaries

Let x_i and y_i be linear processes having the MA(∞) forms

$$x_i = \sum_{j=0}^{\infty} b_j u_{i-j}, \quad y_i = \sum_{l=0}^{\infty} c_l w_{i-l} \tag{4.1}$$

Asymptotics for Fractional Processes. James Davidson, Oxford University Press. © James Davidson (2025).
DOI: 10.1093/9780198955207.003.0004

where $b_j \sim d_x j^{d_x-1}L_x(j)$ and $c_l \sim d_y l^{d_y-1}L_y(l)$. The analysis will initially focus on the following particular setup.

4.1 Assumption The pair (x_i, y_i) are defined by (4.1) where

(a) $|d_x| < \frac{1}{2}$, $|d_y| < \frac{1}{2}$

(b) the sequences (u_i, w_i) are i.i.d. with means of zero, L_r-bounded with $r \geq 2$ and with contemporaneous covariance matrix

$$\mathrm{E}\begin{bmatrix} u_i \\ w_i \end{bmatrix}\begin{bmatrix} u_i & w_i \end{bmatrix} = \boldsymbol{\Sigma} = \begin{bmatrix} \sigma_u^2 & \sigma_{uw} \\ \sigma_{uw} & \sigma_w^2 \end{bmatrix} \tag{4.2}$$

and $\mu_{uw}^4 = \mathrm{E}(u_i^2 w_i^2) < \infty$. □

These random variables define a filtration $\boldsymbol{F} = \{\mathcal{F}_i, i \in \mathbb{Z}\}$ on the probability space such that the processes $\{x_i, \mathcal{F}_i\}$ and $\{y_i, \mathcal{F}_i\}$ are adapted. The assumption of independent shocks is restrictive, but has the benefit of simplicity while focusing interest on the long memory characteristics of the data. The extensions necessary to allow for weak dependence (specifically, autocorrelation) in the shock processes are treated in Chapter 8. The case $u = w$ is generally allowed, so that if in addition $d_x = d_y$ the analysis to follow goes though for the case $x = y$.

The next two results relate to the empirical covariance of a pair of stationary fractional series. Define $\sigma_{xy} = \mathrm{E}(x_i y_i)$. Since under Assumption **4.1**(a) the sequence

$$b_j c_j \sim d_x d_y j^{d_x+d_y-2}L_x(j)L_y(j)$$

is summable and the shocks are serially uncorrelated, the relation

$$\sigma_{xy} = \sigma_{uw}\sum_{j=0}^{\infty} b_j c_j < \infty \tag{4.3}$$

follows in a similar way to (1.5).

4.2 Theorem Under Assumption **4.1**, $n^{-1}\sum_{i=1}^n x_i y_i \to_{L_2} \sigma_{xy}$.

Proof

$$\mathrm{E}\left(\frac{1}{n}\sum_{i=1}^n x_i y_i - \sigma_{xy}\right)^2 = \frac{1}{n^2}\sum_{j=0}^{\infty}\sum_{m=0}^{\infty}\sum_{l=0}^{\infty}\sum_{p=0}^{\infty} b_j b_m c_l c_p$$

$$\times \sum_{i=1}^n\sum_{k=1}^n \left(\mathrm{E}(u_{i-j}w_{i-l}u_{k-m}w_{k-p}) - \mathrm{E}(u_{i-j}w_{i-l})\mathrm{E}(u_{k-m}w_{k-p})\right) \tag{4.4}$$

where Assumption **4.1**(b) implies that

$$\mathrm{E}(u_{i-j}w_{i-l}u_{k-m}w_{k-p}) - \mathrm{E}(u_{i-j}w_{i-l})\mathrm{E}(u_{k-m}w_{k-p})$$
$$= \begin{cases} \mu_{uw}^4 - \sigma_{uw}^2 & i-j = i-l = k-m = k-p \\ \sigma_u^2\sigma_w^2 & i-j = k-m \neq i-l = k-p \\ 0 & \text{otherwise.} \end{cases} \tag{4.5}$$

It has to be shown that the terms of (4.4) that are subject to the two types of restriction in (4.5) are collectively of small order. Writing out the terms in which the first restriction holds, since $j = l$ and $m = p$ the fourfold sum reduces to a twofold sum. Since u and w are always contemporaneous in these terms, they can be combined as $z_i = u_iw_i - \mathrm{E}(u_iw_i)$ and by Assumption **4.1** this series is serially independent with mean zero and variance $\mu_{uw}^4 - \sigma_{uw}^2$. Also, let $h_j = b_jc_j = O(j^{d_x+d_y-2})$. If these substitutions are made in (4.4) the result is

$$\frac{1}{n^2}\sum_{j=0}^{\infty}\sum_{m=0}^{\infty}h_jh_m\sum_{i=1}^{n}\sum_{k=1}^{n}\mathrm{E}(z_{i-j}z_{k-m}) = \frac{1}{n^2}\mathrm{E}\left(\sum_{j=0}^{\infty}\sum_{i=1}^{n}h_jz_{i-j}\right)^2. \tag{4.6}$$

Next, re-order the terms so that the coefficients attached to each z_i are gathered together (compare (2.23)). In view of the serial independence and the summability of $\{h_j\}$, the right-hand side of (4.6) is

$$\frac{1}{n^2}\mathrm{E}\left(\sum_{j=0}^{\infty}\left(\sum_{k=\max\{0,j-n+1\}}^{j}h_k\right)z_{n-j}\right)^2$$
$$= \frac{1}{n^2}\mathrm{E}\left(\sum_{j=0}^{n-1}\left(\sum_{k=0}^{j}h_k\right)z_{n-j} + \sum_{j=n}^{\infty}\left(\sum_{k=j-n+1}^{j}h_k\right)z_{n-j}\right)^2$$
$$= \frac{\mu_{uw}^4 - \sigma_{uw}^2}{n^2}\left(\sum_{j=0}^{n-1}\left(\sum_{k=0}^{j}h_k\right)^2 + \left(\sum_{j=n}^{2n-1} + \sum_{j=2n}^{\infty}\right)\left(\sum_{k=j-n+1}^{j}h_k\right)^2\right)$$
$$= O(n^{-1}). \tag{4.7}$$

Of the three sums appearing in the penultimate member of (4.7), the first two are $O(n)$ and the third is $o(n)$.

The second restriction in (4.5) implies that

$$\mathrm{E}(u_{i-j}w_{i-l}u_{k-m}w_{k-p}) - \mathrm{E}(u_{i-j}w_{i-l})\mathrm{E}(u_{k-m}w_{k-p})$$
$$= \mathrm{E}(u_{i-j}u_{k-m})\mathrm{E}(w_{i-l}w_{k-p}).$$

Excluding the cases where one or other of these covariances is zero implies $m = j+k-i$ and $l = p+i-k$, which require respectively that $j \geq i-k$ and that $p \geq k-i$. Therefore the summands do not depend on i and k independently, but only on $i-k$. Set $q = i-k$, noting that $1-n \leq q \leq n-1$ and not overlooking that in the order of magnitude calculations $0^{d_x-1} = 0^{d_y-1} = 1$, representing b_0 and c_0. The sum has the form

$$\begin{aligned}
&\frac{1}{n^2}\sum_{j=0}^{\infty}\sum_{m=0}^{\infty}\sum_{l=0}^{\infty}\sum_{p=0}^{\infty}\sum_{i=1}^{n}\sum_{k=1}^{n} b_j b_m c_l c_p \mathrm{E}(u_{i-j}u_{k-m})\mathrm{E}(w_{i-l}w_{k-p}) \\
&= \frac{\sigma_u^2\sigma_w^2}{n^2}\sum_{q=1-n}^{n-1}\sum_{j=\max\{0,q\}}^{\infty} b_j b_{j-q} \sum_{p=\max\{0,-q\}}^{\infty} c_p c_{p+q} \\
&\ll \frac{1}{n^2}\sum_{q=1-n}^{n-1}\sum_{j=\max\{0,q\}}^{\infty} j^{2d_x-2}\left(1-\frac{q}{j}\right)^{d_x-1} \sum_{p=\max\{0,-q\}}^{\infty} p^{2d_y-2}\left(1+\frac{q}{p}\right)^{d_y-1} \\
&\ll \frac{1}{n^2}\sum_{q=0}^{n-1}(q^{2d_x-1}+q^{2d_y-1}) \\
&= O(n^{-1}). \qquad \blacksquare
\end{aligned}$$

A key application of this result is to the case $y_i = x_i$, with the limit in (1.5). It is proved under weak dependence of the shocks as Theorem **8.5**.

The next theorem, applying only to cases with $y_i \neq x_i$, deals with the distribution of the empirical covariance when the shocks are contemporaneously as well as serially independent.

4.3 Theorem Under Assumption **4.1**, if $d_x + d_y < \frac{1}{2}$ and shock processes $\{u_i\}$ and $\{w_i\}$ are independently distributed,

$$\frac{1}{\sqrt{n}}\sum_{i=1}^{n} x_i y_i \xrightarrow{d} \mathrm{N}(0, V_{xy})$$

where $V_{xy} < \infty$.

Proof Decompose the sum as

$$\frac{1}{\sqrt{n}}\sum_{i=1}^{n} x_i y_i = \sum_{k=0}^{\infty}\sum_{j=0}^{\infty} b_k c_j \left(\frac{1}{\sqrt{n}}\sum_{i=1}^{n} u_{i-k}w_{i-j}\right).$$

Random variables $Z(k,j)$ defined as the limits

$$\frac{1}{\sqrt{n}}\sum_{i=1}^{n} u_{i-k}w_{i-j} \xrightarrow{\mathrm{d}} Z(k,j) \tag{4.8}$$

as $n \to \infty$ are distributed as $\mathrm{N}(0, \sigma_u^2\sigma_w^2)$ for any j and k, under the assumptions. These limit distributions can be shown by the Lindeberg-Lévy central limit theorem or equivalent. It can be verified that $Z(k,j) = Z(k',j')$ if $j - k = j' - k'$. Moreover, products of the form $u_{i-k}w_{i-j}u_{m-k'}w_{m-j'}$ have zero mean unless both $i - k = m - k'$ and $i - j = m - j'$, which imposes the same condition. In other words, if $j - j' \neq k - k'$ then $\mathrm{E}(Z(k,j)Z(k',j')) = 0$. Being Gaussian, the pairs are independent if they are not identical.

Define $\zeta = \sum_{k=0}^{\infty}\sum_{j=0}^{\infty} b_k c_j Z(k,j)$. Being a weighted sum of Gaussian random variables that are either identical or independent, ζ is itself Gaussian if its variance is finite. The pairs $b_k c_j Z(k,j)$ and $b_{k'}c_{j'}Z(k',j')$ are correlated if, for any $p \in \mathbb{N}$, both $k' = k + p$ and $j' = j + p$ and also if both $k = k' + p$ and $j = j' + p$. Summing the corresponding weights, the variance is calculated as

$$V_{xy} = \mathrm{E}(\zeta^2) = \sigma_u^2\sigma_w^2\left(\sum_{k=0}^{\infty} b_k^2 \sum_{j=0}^{\infty} c_j^2 + 2\sum_{p=1}^{\infty}\left(\sum_{k=0}^{\infty} b_k b_{k+p}\sum_{j=0}^{\infty} c_j c_{j+p}\right)\right). \tag{4.9}$$

By (1.6), $\sum_{k=0}^{\infty} b_k b_{k+p}\sum_{j=0}^{\infty} c_j c_{j+p} = O(p^{2d_x+2d_y-2})$ and these terms are summable over p under the specified assumption. ∎

In fact there is no necessity for u and w to be independent in this result. If the products have zero mean and finite variance, serial independence is more than sufficient for the central limit theorem to operate. The corresponding result under weak dependence is proved in §8.1 as Theorem **8.6**.

Some interesting implications of the condition $d_x + d_y < \frac{1}{2}$ are discussed in §7.3. A natural example is the case where x exhibits stationary long memory and y is weakly dependent, with $d_y = 0$ so that $c_j = 0$ for $j > 0$. (This means serially independent under Assumption **4.1**.) In this case, simply enough, $V_{xy} = \sigma_w^2\sigma_u^2\sum_{k=0}^{\infty} b_k^2$. However, a limited degree of long memory in both processes is also compatible with a Gaussian limit.

4.2 The Covariance Decomposition

In the remainder of this chapter, and also in Chapters 5–6 to follow, the case of principal interest is going to be the limiting distribution of the covariance process

$$G_n = \frac{1}{nK(n)} \sum_{i=1}^{n-1} \sum_{k=1}^{i} x_k y_{i+1} \tag{4.10}$$

where

$$K(n) = n^{d_x+d_y} L_x(n) L_y(n). \tag{4.11}$$

An equivalent notation would be $G_n = (nK(n))^{-1} \sum_{i=1}^{n-1} S_i y_{i+1}$ where S_i denotes the partial sum process. Strictly, it would be better to write (4.10) as G_n^{xy} so as to distinguish the case in which the roles of x and y are interchanged. This form of labelling becomes unavoidable in some later developments, but to avoid clutter the distinction is treated as implicit where this does not hinder sense.

The natural context for statistics such as (4.10) is regression analysis and specifically cointegrating regression. These applications are reviewed in Chapter 7. The aim is to show that the weak limit of G_n exists and for fBms X and Y is a random variable $\int_0^1 X\mathrm{d}Y$, with known distribution. However, $\int_0^1 X\mathrm{d}Y$ here defined is not in general an Itô integral and in particular it does not have a mean of zero.

Serial uncorrelatedness of the shocks allows G_n in (4.10) to be split into components according to their expected values. Substituting from (4.1), the sum of products may be expanded as

$$G_n = \frac{1}{nK(n)} \sum_{i=1}^{n-1} \sum_{k=1}^{i} \sum_{j=0}^{\infty} \sum_{l=0}^{\infty} b_j c_l u_{k-j} w_{i+1-l}. \tag{4.12}$$

Let this sum be decomposed as $G_n = G_{1n} + G_{2n} + G_{3n}$ where G_{1n} contains those terms in which $k - j \leq i - l$, so that the time indices of w strictly exceed those of u, and in G_{3n} the indices of u lead those of w, so that $k - j > i + 1 - l$. Therefore, G_{2n} has the terms where $k - j = i + 1 - l$ such that the time indices of u and w match. The decomposition can be written out as

$$G_{1n} = \frac{1}{nK(n)} \sum_{i=1}^{n-1} \sum_{k=1}^{i} \sum_{j=0}^{\infty} \sum_{l=0}^{i+j-k} b_j c_l u_{k-j} w_{i+1-l} \tag{4.13}$$

$$G_{2n} = \frac{1}{nK(n)} \sum_{i=1}^{n-1} \sum_{k=1}^{i} \sum_{j=0}^{\infty} b_j c_{i+j-k+1} u_{k-j} w_{k-j} \tag{4.14}$$

and

$$G_{3n} = \frac{1}{nK(n)} \sum_{i=1}^{n-1} \sum_{k=1}^{i} \sum_{j=0}^{\infty} \sum_{l=i+j-k+2}^{\infty} b_j c_l u_{k-j} w_{i+1-l}. \tag{4.15}$$

With $\mathrm{E}(G_{1n}) = \mathrm{E}(G_{3n}) = 0$ under Assumption **4.1**, the first objective is to find a formula for the limiting mean of G_{2n}.

4.4 Theorem If Assumption **4.1** holds and $d_x + d_y > 0$, $\mathrm{E}(G_{2n}) \to \sigma_{uw}\lambda_{xy}$ as $n \to \infty$ where

$$\lambda_{xy} = \frac{1}{d_x + d_y}\left(\frac{d_y}{1 + d_x + d_y} + \int_0^\infty \left(d_y(1+\tau)^{d_x+d_y} + d_x\tau^{d_x+d_y} - (d_x + d_y)(1+\tau)^{d_y}\tau^{d_x}\right)\mathrm{d}\tau\right). \quad (4.16)$$

Proof It can be verified by inspection that

$$\sum_{i=1}^{n-1}\sum_{k=1}^{i}\sum_{j=0}^{\infty} b_j c_{i+j-k+1} = \sum_{i=1}^{n-1}(n-i)\sum_{j=0}^{\infty} b_j c_{j+i} = \sum_{i=1}^{n-1}\sum_{k=0}^{\infty}\sum_{j=\max\{0,k-i+1\}}^{k} b_j c_{k+1}.$$

The sum of the b_j in the right-hand member is equal to $a_{n,i-k}(i/n, 0)$ as defined in (2.26), a sum containing either $k + 1$ terms or i terms, whichever is the smaller. Therefore, setting $\mathrm{E}(u_{k-j}w_{k-j}) = \sigma_{uw}$ in (4.14) it is found that

$$\mathrm{E}(G_{2n}) = \frac{\sigma_{uw}}{nK(n)}\sum_{i=1}^{n-1}\sum_{k=0}^{\infty} a_{n,i-k}(i/n, 0)c_{k+1}. \quad (4.17)$$

To obtain the limit of this sum, separate the terms into two blocks, for $k = 0, \ldots, i - 1$ and $k \geq i$ respectively. Applying (2.30) and (1.2) while noting that the slowly varying components cancel in the limit and that $d_x + d_y > 0$ by assumption, the first block gives

$$\begin{aligned}\frac{\sigma_{uw}}{nK(n)}\sum_{i=1}^{n-1}\sum_{k=0}^{i-1} a_{n,i-k}(i/n, 0)c_{k+1} &\sim \frac{\sigma_{uw}d_y}{n^2}\sum_{i=1}^{n-1}\sum_{k=0}^{i-1}\left(\frac{k+1}{n}\right)^{d_x+d_y-1} \\ &\to \sigma_{uw}d_y\int_0^1\int_0^\tau \xi^{d_x+d_y-1}\mathrm{d}\xi\mathrm{d}\tau \\ &= \frac{\sigma_{uw}d_y}{(d_y + d_x)(1 + d_y + d_x)}. \end{aligned} \quad (4.18)$$

Applying (2.31), the second block gives

$$\begin{aligned}
&\frac{\sigma_{uw}}{nK(n)}\sum_{i=1}^{n-1}\sum_{k=i}^{\infty}a_{n,i-k}(i/n,0)c_{k+1} \\
&\quad\sim \frac{\sigma_{uw}d_y}{n^2}\sum_{i=1}^{n-1}\sum_{k=0}^{\infty}\left(\left(\frac{i+k}{n}\right)^{d_x}-\left(\frac{k}{n}\right)^{d_x}\right)\left(\frac{i+k}{n}\right)^{d_y-1} \\
&\quad\to \sigma_{uw}d_y\int_0^{\infty}\int_0^1((\xi+\tau)^{d_x}-\tau^{d_x})(\xi+\tau)^{d_y-1}\mathrm{d}\xi\mathrm{d}\tau \\
&\quad= \frac{\sigma_{uw}}{(d_x+d_y)}\int_0^{\infty}\Big(d_y(1+\tau)^{d_x+d_y}+d_x\tau^{d_x+d_y} \\
&\qquad\qquad -(d_x+d_y)(1+\tau)^{d_y}\tau^{d_x}\Big)\mathrm{d}\tau. \qquad (4.19)
\end{aligned}$$

Combining these two limits completes the proof. ∎

If $d_y = 0$ then $\lambda_{xy} = 0$. One way this might arise is because $c_k = 0$ for $k > 0$, making y_i an independent process under Assumption **1.1**, but the assumption $c_k = O(k^{-1-\delta})$ for $\delta > 0$ is asymptotically equivalent to independence provided that $\sum_{k=0}^{\infty} c_k \neq 0$, ruling out antipersistence. This latter case is obtained by setting $d_y = -\delta$ and also $K(n) = n^{d_x}L_x(n)$, from which $\mathrm{E}(G_{2n}) = O(n^{-\delta})$ must follow.

However, for (4.16) to make sense, the inequality $d_x + d_y > 0$ must be strict. To see what is going on here, an illuminating step is to let λ_{yx} represent the counterpart of (4.16) with the roles of x and y interchanged, so that d_x and d_y are swapped in the formula. Then $\lambda_{xy} + \lambda_{yx} = \Upsilon_{xy}$, where it is easily verified that

$$\Upsilon_{xy} = \frac{1}{1+d_x+d_y} + \int_0^{\infty}\left((1+\tau)^{d_x}-\tau^{d_x}\right)\left((1+\tau)^{d_y}-\tau^{d_y}\right)\mathrm{d}\tau \qquad (4.20)$$

and, needless to say, $\Upsilon_{yx} = \Upsilon_{xy}$. This expression exists whether or not $d_x + d_y > 0$ and if $d_x = d_y = 0$ then $\Upsilon_{xy} = 1$. Υ_{xy} is the bivariate case of Υ_{kl} in (2.50), the off-diagonal element of the long-run covariance matrix of the process (x_i, y_i), such that

$$\sigma_{uw}\Upsilon_{xy} = \lim_{n\to\infty}\frac{1}{nK(n)}\mathrm{E}\left(\sum_{i=1}^{n}x_i\sum_{i=1}^{n}y_i\right). \qquad (4.21)$$

A natural case to consider is $y = x$ for which, provided $d_x > 0$, it is found that $\lambda_{xx} = \frac{1}{2}\Upsilon_{xx}$ and also that $\Upsilon_{xx} = \Upsilon_{d_x}$, this being the variance of the univariate case from (2.4).

This last example draws attention to the absence from the formula of the contemporaneous variance, σ_x^2 defined in (1.5), and more generally of the contemporaneous covariance σ_{xy} defined in (4.3). An explanation is forthcoming from the decomposition of the product of sums whose limiting expectation appears in (4.21). Multiplying out gives

$$\sum_{i=1}^{n} x_i \sum_{i=1}^{n} y_i = \sum_{i=1}^{n} x_i y_i + \sum_{i=1}^{n-1}\sum_{k=1}^{i} x_k y_{i+1} + \sum_{i=1}^{n-1}\sum_{k=1}^{i} y_k x_{i+1}. \tag{4.22}$$

The expectation of the first right-hand-side term in (4.22) is $n\sigma_{xy}$ where σ_{xy} is finite, whereas after dividing by $nK(n)$ the expectations of the second and third terms converge respectively to $\sigma_{uw}\lambda_{xy}$ and $\sigma_{uw}\lambda_{yx}$. With $d_x + d_y > 0$ the contemporaneous component of the covariance therefore vanishes under the normalization by $nK(n)$, which explains why Υ_{xy} can be decomposed into just two complementary terms. This is a key distinction between the distributions of long memory processes and the weakly dependent case.

4.3 Closed Forms

Closed forms for the expressions in (4.16) and (4.20) can be derived by arguments similar to those of Lemma **2.3**. The following proof is adapted from Proposition 3.2 of [20].

4.5 Theorem Under Assumption **4.1**(a) and if $d_x + d_y > 0$,

$$\lambda_{xy} = \Gamma(d_x+1)\Gamma(d_y+1)\frac{\Gamma(1-d_x-d_y)\sin\pi d_y}{\pi\left(1+d_x+d_y\right)\left(d_x+d_y\right)}. \tag{4.23}$$

Proof The task is to evaluate the integral in the second term of (4.16). Write the integrand, with a complex argument, as

$$\begin{aligned}\phi_{xy}(z) &= d_y(1+z)^{d_x+d_y} + d_x z^{d_x+d_y} - (d_x+d_y)(1+z)^{d_y} z^{d_x} \\ &= d_y(1+z)^{d_y}\left((1+z)^{d_x} - z^{d_x}\right) - d_x z^{d_x}\left((1+z)^{d_y} - z^{d_y}\right)\end{aligned} \tag{4.24}$$

for $z \in \mathbb{C}$. Defining

$$z(\xi) = \begin{cases} \xi, & \xi \geq 0 \\ e^{i\pi}|\xi|, & \xi < 0 \end{cases}$$

similarly to (2.8), $f_{xy} = \phi_{xy} \circ z$ is a well-defined function of a real variable $\xi \in \mathbb{R}$. The integral appearing in (4.16), whose evaluation is the objective, can then be written

$$\mathcal{L}_{xy} = \int_0^\infty f_{xy}(\xi)\mathrm{d}\xi. \tag{4.25}$$

In (4.24), as z increases $(1+z)^{d_x} - z^{d_x}$ approaches $d_x z^{d_x-1}$ and $(1+z)^{d_y} - z^{d_y}$ approaches $d_y z^{d_y-1}$, so that $f_{xy}(\xi) = O(\xi^{d_x+d_y-2})$ as $\xi \to \pm\infty$. Hence, f_{xy} is integrable if $d_x > 0$ and $d_y > 0$. If $d_x < 0$, f_{xy} possesses a singularity at $\xi = 0$ but $f_{xy}(\xi)/z(\xi)^{d_x} \to -(d_x + d_y)$ as $\xi \to 0$. Similarly, if $d_y < 0$ then f_{xy} possesses a singularity at $\xi = -1$, but $f_{xy}(\xi)/(1 + z(\xi))^{d_y} \to -(d_x + d_y)\mathrm{e}^{\mathrm{i}\pi d_x}$ as $\xi \to -1$. Therefore f_{xy} is integrable for all $\xi \in \mathbb{R}$.

Let f_{yx} denote the formula complementary to f_{xy} with y and x interchanged. Introduce the change of variable $\tau = -1 - \xi$, so that $\xi = -1 - \tau$. Then, similarly to (2.9), $z(\tau) = \mathrm{e}^{\mathrm{i}\pi}(1 + z(\xi))$ and $1 + z(\tau) = \mathrm{e}^{\mathrm{i}\pi}z(\xi)$. These relations imply according to (4.24) that

$$f_{yx}(\tau) = \mathrm{e}^{\mathrm{i}\pi(d_x+d_y)}f_{xy}(\xi) \tag{4.26}$$

and by symmetry of the formulae, $f_{xy}(\xi) = \mathrm{e}^{\mathrm{i}\pi(d_x+d_y)}f_{yx}(\tau)$ likewise. It follows that $f_{xy}(\xi) = \mathrm{e}^{2\mathrm{i}\pi(d_x+d_y)}f_{xy}(\xi)$ and hence that

$$\int_{-\infty}^\infty f_{xy}(\xi)\mathrm{d}\xi = 0 \tag{4.27}$$

and equally that

$$\int_{-\infty}^\infty f_{yx}(\tau)\mathrm{d}\tau = 0. \tag{4.28}$$

Now integrate f_{yx} over the three regions of the line, $(0,\infty)$, $(-1,0)$, and $(-\infty,-1)$, knowing from (4.28) that the sum of the components is zero. The first of these components is the case complementary to (4.25),

$$\mathcal{L}_{yx} = \int_0^\infty f_{yx}(\tau)\mathrm{d}\tau. \tag{4.29}$$

With $\tau = -\xi - 1$, in view of (4.26) the third component of (4.28) is

$$\int_{-\infty}^{-1} f_{yx}(\tau)\mathrm{d}\tau = \int_0^\infty f_{yx}(\xi)\mathrm{d}\xi = \mathrm{e}^{\mathrm{i}\pi(d_x+d_y)}\mathcal{L}_{xy}. \tag{4.30}$$

Lastly, the middle component rearranges in a manner similar to (2.10) as

$$\int_{-1}^{0} f_{yx}(\tau)\mathrm{d}\tau = -d_x \int_0^1 (1-\tau)^{d_x+d_y}\mathrm{d}\tau + d_y \int_0^1 (\mathrm{e}^{\mathrm{i}\pi}\tau)^{d_x+d_y}\mathrm{d}\tau$$

$$- (d_x + d_y) \int_0^1 (1-\tau)^{d_x}(\mathrm{e}^{\mathrm{i}\pi}\tau)^{d_y}\mathrm{d}\tau$$

$$= \frac{d_x + \mathrm{e}^{\mathrm{i}\pi(d_x+d_y)}d_y}{1+d_x+d_y} - (d_x+d_y)\mathrm{e}^{\mathrm{i}\pi d_y}\int_0^1 (1-\tau)^{d_x}\tau^{d_y}\mathrm{d}\tau. \tag{4.31}$$

Adding together (4.29), (4.30), and (4.31), applying (B.14) yields

$$\mathcal{L}_{yx} + \frac{d_x + \mathrm{e}^{\mathrm{i}\pi(d_x+d_y)}d_y}{1+d_x+d_y} - (d_x+d_y)\mathrm{e}^{\mathrm{i}\pi d_y}B(d_x+1, d_y+1) + \mathrm{e}^{\mathrm{i}\pi(d_x+d_y)}\mathcal{L}_{xy} = 0. \tag{4.32}$$

This equality remains true if x and y are everywhere interchanged, the Beta function being symmetric in its arguments. Thus,

$$\mathcal{L}_{xy} + \frac{d_y + \mathrm{e}^{\mathrm{i}\pi(d_y+d_x)}d_x}{1+d_y+d_x} - (d_y+d_x)\mathrm{e}^{\mathrm{i}\pi d_x}B(d_y+1, d_x+1) + \mathrm{e}^{\mathrm{i}\pi(d_y+d_x)}\mathcal{L}_{yx} = 0. \tag{4.33}$$

Subtract (4.32) multiplied by $\mathrm{e}^{\mathrm{i}\pi(d_x+d_y)}$ from (4.33), so eliminating $\mathcal{L}_{yx}$. After cancellation and rearrangement this operation gives the integral in (4.25) as

$$\mathcal{L}_{xy} = \frac{\mathrm{e}^{\mathrm{i}\pi d_x}(1-\mathrm{e}^{2\mathrm{i}\pi d_y})}{1-\mathrm{e}^{2\mathrm{i}\pi(d_x+d_y)}}(d_x+d_y)B\left(d_x+1, d_y+1\right) - \frac{d_y}{1+d_x+d_y}. \tag{4.34}$$

To simplify this expression, identities (B.2), (B.9), (B.11), and (B.15) successively give

$$\frac{\mathrm{e}^{\mathrm{i}\pi d_x}(1-\mathrm{e}^{2\mathrm{i}\pi d_y})}{1-\mathrm{e}^{2\mathrm{i}\pi(d_x+d_y)}} = \frac{\mathrm{e}^{\mathrm{i}\pi d_x}(1-\mathrm{e}^{2\mathrm{i}\pi d_y})(1-\mathrm{e}^{-2\mathrm{i}\pi(d_x+d_y)})}{2(1-\cos 2\pi(d_x+d_y))}$$

$$= \frac{2\cos\pi d_x - 2\cos\pi(d_x+2d_y)}{4\sin^2\pi(d_x+d_y)}$$

$$= \frac{\sin\pi d_y}{\sin\pi\left(d_x+d_y\right)}$$

$$= \frac{\Gamma(d_x+d_y)\Gamma(1-d_x-d_y)\sin\pi d_y}{\pi}.$$

Also, by (B.14) and (B.13),

$$B\left(d_x+1, d_y+1\right) = \frac{\Gamma(d_x+1)\Gamma(d_y+1)}{(d_x+d_y)(1+d_x+d_y)\Gamma(d_x+d_y)}$$

Substituting these formulae into (4.34) and simplifying gives

$$\mathcal{L}_{xy} = \Gamma(d_x+1)\Gamma(d_y+1)\frac{\Gamma(1-d_x-d_y)\sin\pi d_y}{\pi(1+d_x+d_y)} - \frac{d_y}{1+d_x+d_y}.$$

Finally, substituting this expression into (4.16) gives (4.23). ∎

Noticing how λ_{yx} is obtained from (4.23) by changing $\sin\pi d_y$ to $\sin\pi d_x$, it follows directly that

$$\begin{aligned}\Upsilon_{xy} &= \lambda_{xy} + \lambda_{yx} \\ &= \Gamma(d_x+1)\Gamma(d_y+1)\frac{\Gamma(1-d_x-d_y)(\sin\pi d_y + \sin\pi d_x)}{\pi\left(1+d_x+d_y\right)\left(d_x+d_y\right)}. \end{aligned} \quad (4.35)$$

However, formula (4.35) has been derived on the assumption $d_x + d_y > 0$ and returns an awkward 'zero over zero' result for the case $d_x = -d_y$, even though expression (4.20) is well defined for $d_x + d_y \leq 0$. There is an equivalent and more robust version of the closed formula that has already been quoted as expression (2.54), in connection with Theorem **2.9**. The equivalence between the two representations is shown as follows.

4.6 Theorem

$$\Upsilon_{xy} = \frac{\Gamma(d_x+1)\Gamma(d_y+1)\cos(\pi(d_x-d_y)/2)}{\Gamma(d_x+d_y+2)\cos(\pi(d_x+d_y)/2)}. \quad (4.36)$$

Proof Applying successively identities (B.8), (B.10), and (B.15) gives the relation

$$\begin{aligned}\sin\pi d_y + \sin\pi d_x &= \frac{\sin\pi(d_x+d_y)(\sin\pi d_y + \sin\pi d_x)}{2\sin(\pi(d_x+d_y)/2)\cos(\pi(d_x+d_y)/2)} \\ &= \sin\pi(d_x+d_y)\frac{\cos(\pi(d_x-d_y)/2)}{\cos(\pi(d_x+d_y)/2)} \\ &= \frac{\pi}{\Gamma(d_x+d_y)\Gamma(1-d_x-d_y)}\frac{\cos(\pi(d_x-d_y)/2)}{\cos(\pi(d_x+d_y)/2)}. \end{aligned} \quad (4.37)$$

Substituting (4.37) into (4.35), replacing $\Gamma(d_x + d_y)\left(1 + d_x + d_y\right)(d_x + d_y)$ by $\Gamma(d_x + d_y + 2)$ according to (B.13) and then cancelling the matching terms in the ratio gives finally (4.36). ∎

A direct derivation of this result based on the harmonizable representation of the processes is Theorem **9.3**.

4.4 Antipersistence

The condition $d_x + d_y > 0$ must hold to define λ_{xy} and λ_{yx}, without which at least one of the processes is antipersistent. The following results give an indication of what happens in this latter case. The cases $d_x + d_y < 0$ and $d_x + d_y = 0$ require separate treatment, while $d_x = -d_y \neq 0$ must also be distinguished from $d_x = d_y = 0$, which is the conventional unit root scenario.

4.7 Theorem

(i) If $d_x = -d_y \neq 0$ and $\sigma_{uw} \neq 0$ then $\mathrm{E}(G_{2n}) = O(\log n)$.

(ii) If $d_x + d_y < 0$ and $\sigma_{uw} \neq 0$ then $\mathrm{E}(G_{2n}) = O(n^{-d_x-d_y})$.

Proof For part (i), consider expression (4.17) and its decomposition into the terms with $k < i$ in (4.18) and those with $k \geq i$ in (4.19). With $d_x + d_y = 0$ the second member of (4.18) takes the form

$$\frac{\sigma_{uw}d_y}{n}\sum_{i=1}^{n-1}\sum_{k=1}^{i}k^{-1} = O\left(\frac{1}{n}\sum_{i=1}^{n-1}\log i\right) = O(\log n) \tag{4.38}$$

where the second equality applies Stirling's approximation (B.16). For the terms with $k \geq i$, consider the second member of (4.19). Applying a Taylor expansion of first order around $(i + k)/n$ gives, with $d_x + d_y = 0$,

$$\begin{aligned}
&\frac{\sigma_{uw}d_y}{n^2}\sum_{i=1}^{n-1}\sum_{k=0}^{\infty}\left(\left(\frac{i+k}{n}\right)^{d_x} - \left(\frac{k}{n}\right)^{d_x}\right)\left(\frac{i+k}{n}\right)^{d_y-1} \\
&\qquad \sim \frac{\sigma_{uw}d_yd_x}{n^2}\sum_{i=1}^{n-1}\sum_{k=0}^{\infty}\left(\frac{i+k}{n}\right)^{-2}\frac{i}{n} \\
&\qquad = \sigma_{uw}d_yd_x\sum_{i=1}^{n-1}i^{-1}\left(\frac{1}{n}\sum_{k=0}^{\infty}\left(\frac{i}{i+k}\right)^2\right) = O(\log n).
\end{aligned} \tag{4.39}$$

The indicated order of magnitude in (4.39) is verified by showing that the term in parentheses in the penultimate member is $O(1)$ as $i \to n$ and $n \to \infty$.

Since $i < n$,

$$\frac{1}{n}\sum_{k=0}^{\infty}\left(\frac{i}{i+k}\right)^2 \leq \sum_{k=0}^{\infty}k^{-1-\delta}\left(\frac{i^{1/2}k^{(1+\delta)/2}}{i+k}\right)^2$$

where $0 < \delta < 1$. The squared term in parentheses converges to zero as $k \to \infty$ for any fixed i, and also as $i \to \infty$.

For part (ii), the condition $d_x + d_y < 0$ means that for each $i < n$ the terms $a_{n,i-k}(i/n, 0)c_{k+1} = O(k^{d_x+d_y-1})$ are summable over k. In view of (2.26), these sums have the form

$$\sum_{k=0}^{\infty} a_{n,i-k}(i/n,0)c_{k+1} = \sum_{k=0}^{i-1}\left(\sum_{j=0}^{k} b_j\right)c_{k+1} + \sum_{k=i}^{\infty}\left(\sum_{j=k-i+1}^{k} b_j\right)c_{k+1}. \tag{4.40}$$

The second block of terms in (4.40) vanishes as $i \to \infty$ and the first block has a finite limit as $i \to \infty$. Given $b_0 = 1$ it does not vanish unless $d_y = 0$ and $c_{k+1} = 0$ for $k \geq 0$, but it is at most $O(1)$. The conclusion is that

$$\frac{1}{n}\sum_{i=1}^{n-1}\sum_{k=0}^{\infty} a_{n,i-k}(i/n,0)c_{k+1} = O(1)$$

and the result follows in view of the normalization $nK(n)$ defining G_n in (4.12). ∎

In the case $d_x + d_y < 0$, the summability means that all three terms of the decomposition (4.22) are $O_p(n)$. It can generally be assumed that there exists a finite constant

$$\gamma_{xy} = \lim_{n\to\infty}\frac{1}{n}\sum_{i=1}^{n-1}\sum_{k=1}^{i} \mathrm{E}(x_k y_{i+1}) \tag{4.41}$$

and a complementary quantity γ_{yx}. However, there are no general formulae for γ_{xy} and γ_{yx} to compare with $\sigma_{uw}\lambda_{xy}$ and $\sigma_{uw}\lambda_{yx}$ because due to the summability of the terms in (4.41) they must depend on low-order lag coefficients, which are not restricted under (1.2). The one thing known is that $\mathrm{E}(\sum_{i=1}^{n} x_i)^2 = O(n^{2d_x+1})$ and $\mathrm{E}(\sum_{i=1}^{n} y_i)^2 = O(n^{2d_y+1})$ according to Corollary **2.7**, so with $d_x + d_y < 0$ the expected left-hand side of (4.22) is $o(n)$ by the Cauchy-Schwarz inequality. In view of the fact that the expectation of the first right-hand-side term of (4.22) is $O(n)$ by Theorem **4.2**, it must be the case that

$$\sigma_{xy} + \gamma_{xy} + \gamma_{yx} = 0. \tag{4.42}$$

The connection between this equality and the limit in (4.21) where Υ_{xy} is well defined for $d_x + d_y \leq 0$ is the choice of normalization. Under the normalization by n the covariance vanishes, but the limit in (4.21) exists under normalization by $nK(n)$, as was shown in Theorem **2.9** for a pair $d_k = d_x$ and $d_l = d_y$.

Another way to understand (4.42) is as the multivariate counterpart of the property of antipersistent processes noted in §1.4 for the univariate case. If $y = x$ so that necessarily $d_x < 0$, then $\gamma_{xx} = \sum_{k=1}^{\infty} \gamma_k$ where γ_k is defined for x in (1.6). The relation in (4.42) is then identical to that represented in equation (1.21). For the fractionally differenced model represented by (1.13) with $d_x < 0$, the γ_k are negative for every $k \geq 1$ according to (1.16) and in view of (4.42) it follows from (1.18) that

$$\gamma_{xx} = -\frac{\sigma_u^2 \Gamma(1 - 2d_x)}{2\Gamma(1 - d_x)^2}.$$

The condition $\gamma_{xx} < 0$ reflects the fact that the limiting partial sum process (that is to say, X defined in (2.1)) 'reverts to the centre line' due to independence of initial conditions. Its increments are negatively correlated and with a tendency towards zero when $1 + d_x < 1$. In the hierarchy of time dependence, the partial sum process with antipersistent increments falls short of the random walk process ($d_x = 0$) where the direction of travel is, as is well known, independent of position.

The case $d_x = -d_y \neq 0$ involves a slight modification of these conclusions. The interesting feature is that the restriction is an equality instead of a strict inequality, which provides a role for a slowly varying factor in the limit. According to Theorem **4.7**(i), to have $|\gamma_{xy}| < \infty$ it must be the case that

$$\gamma_{xy} = \lim_{n \to \infty} \frac{1}{n \log n} \sum_{i=1}^{n-1} \sum_{k=1}^{i} \mathrm{E}(x_k y_{i+1}). \tag{4.43}$$

Nothing specific can be said about the form of this limit any more than about (4.41), so there is no need to coin a different symbol for it. The main consequence of the changed normalization is that equation (4.22) must be divided by $n \log n$ to avoid divergence and in view of Theorem **4.2**, this has to imply the result

$$\gamma_{xy} + \gamma_{yx} = 0. \tag{4.44}$$

Compare this with the fact that λ_{xy} in (4.23) has the sign of d_y which may differ from that of d_x and hence of λ_{yx}, but there can be no counterpart of (4.44) with $d_x + d_y > 0$.

Don't overlook, finally, that all these conclusions are of interest only because of the assumption $\sigma_{uw} \neq 0$. If u_i and w_i are contemporaneously uncorrelated (implying under Assumption **4.1** that the cross-correlogram is zero at all orders) then of course each of the terms in (4.22) has zero expectation.

4.5 L_2 Convergence

An important fact is that G_{2n} is a consistent estimator of its mean, albeit not a feasible one. The following result implies that the limit distribution of $G_{1n} + G_{3n}$ matches that of the mean deviation of G_n, not overlooking that the mean diverges under the given normalization when $d_x + d_y < 0$ according to Theorem **4.7**. The present assumption of independent shocks simplifies the argument somewhat. The extension to the weak dependence case is given in Chapter 8, as Theorem **8.8**.

4.8 Theorem Under Assumption **4.1**, $\mathrm{E}(G_{2n} - \mathrm{E}(G_{2n}))^2 = O(n^{-1})$.

Proof From (4.14),

$$G_{2n} - \mathrm{E}(G_{2n}) = \frac{1}{nK(n)} \sum_{i=1}^{n-1} \sum_{k=1}^{i} P_{ik} \tag{4.45}$$

where

$$P_{ik} = \sum_{j=0}^{\infty} b_j c_{i+1-k+j}(u_{k-j} w_{k-j} - \sigma_{uw}). \tag{4.46}$$

The square of $nK(n)(G_{2n} - \mathrm{E}(G_{2n}))$ is the summed elements of the outer product of the $\frac{1}{2}n(n-1)$-vector having elements $\{P_{ik}, k = 1, \ldots i, i = 1, \ldots, n-1\}$. It can be verified that

$$\mathrm{E}(G_{2n} - \mathrm{E}(G_{2n}))^2 \leq \frac{2}{n^2 K(n)^2} \sum_{i=1}^{n-1} \sum_{k=1}^{i} \sum_{m=0}^{i-1} \sum_{p=k-i+m}^{k-1} |\mathrm{E}(P_{ik} P_{i-m,k-p})|. \tag{4.47}$$

If i and k in this sum count the rows of the matrix lower triangle, the indices $i - m$ and $k - p$ in the majorant of (4.47) fill in the column elements, with $i - m$ running from 1 to i and $k - p$ running from 1 to $i - m$. Note that the index p can take either sign subject to the inequality $i - m \geq k - p$. There is some double-counting of terms having $m = 0$ in the majorant of (4.47), but to show that it is of small order in n suffices for the proof.

The terms of the sum have a general bound of the form

$$|\mathrm{E}(P_{ik}P_{i-m,k-p})| \leq \sum_{j=0}^{\infty}\sum_{l=0}^{\infty} |b_j b_l c_{i+1-k+j} c_{i-m+1-k+p+l}$$

$$\times \mathrm{E}(u_{k-j}u_{k-p-l}w_{k-j}w_{k-p-l} - \sigma_{uw}^2)| . \qquad (4.48)$$

Under serial independence of the shocks, the expectations in (4.48) vanish unless $j = p + l$ and so this expression reduces to

$$|\mathrm{E}(P_{ik}P_{i-m,k-p})| \leq |\mu_{uw}^4 - \sigma_{uw}^2| \sum_{j=0}^{\infty} |b_j b_{j-p} c_{i+1-k+j} c_{i-m+1-k+j}| . \qquad (4.49)$$

To bound (4.47), the first step is to divide the sum into two components according to the sign of p. In other words, write

$$\mathrm{E}(G_{2n} - \mathrm{E}(G_{2n}))^2 \leq A_{1n} + A_{2n} \qquad (4.50)$$

where A_{1n} has the terms in which $0 \leq p \leq k - 1$ in the innermost sum of (4.47), while A_{2n} contains the remaining terms for the values of m having $k - i + m \leq p < 0$, where these exist. Let A_{1n} be further decomposed by splitting the sum in (4.49) into the sums of the first k terms and of the rest so as to write for the case $p \geq 0$ (noting $b_{j-p} = 0$ when $j < p$),

$$|\mathrm{E}(P_{ik}P_{i-m,k-p})| \leq |\mu_{uw}^4 - \sigma_{uw}^2| \left(\sum_{j=p}^{k-1} + \sum_{j=k}^{\infty} \right) |b_j b_{j-p} c_{i+1-k+j} c_{i-m+1-k+j}|$$

$$= B_{11} + B_{12}. \qquad (4.51)$$

This gives the decomposition

$$A_{1n} = \frac{2}{n^2 K(n)^2} \sum_{i=1}^{n-1} \sum_{k=1}^{i} \sum_{m=0}^{i-1} \sum_{p=\max\{0,k-i+m\}}^{k-1} (B_{11} + B_{12}) = A_{11n} + A_{12n}. \qquad (4.52)$$

Similarly for A_{2n}, for those cases of (4.49) contributing to the sum with $k < i - m$ and hence $p < 0$, decompose the sum as

$$|\mathrm{E}(P_{ik}P_{i-m,k-p})|$$

$$= |\mu_{uw}^4 - \sigma_{uw}^2| \left(\sum_{j=0}^{i-m-k} + \sum_{j=i-m-k+1}^{\infty} \right) |b_j b_{j-p} c_{i+1-k+j} c_{i-m+1-k+j}|$$

$$= B_{21} + B_{22}. \qquad (4.53)$$

The sum over m in (4.47) in this case takes the upper limit of $i-k-1$, so that

$$A_{2n} = \frac{2}{n^2K(n)^2}\sum_{i=1}^{n-1}\sum_{k=1}^{i}\sum_{m=0}^{i-k-1}\sum_{p=k-i+m}^{-1}(B_{21}+B_{22}) = A_{21n} + A_{22n}. \tag{4.54}$$

The terms with $k = i$ are empty here and are assigned the value 0.

The argument now proceeds by bounding each of the terms $A_{11n}, A_{12n}, A_{21n}$, and A_{22n} as functions of n. Substituting from (1.2) but omitting slowly varying components, which will be cancelled in the limit of (4.52), also assigning the value $b_0 = 1$ to 0^{d_x-1} as usual,

$$\begin{aligned} B_{11} &\ll \sum_{j=p}^{k-1} j^{d_x+2d_y-3}(j-p)^{d_x-1}\left(1+\frac{i+1-k}{j}\right)^{d_y-1} \\ &\qquad\qquad \times\left(1+\frac{i-m+1-k}{j}\right)^{d_y-1} \\ &\ll (k-p)^{2d_x+2d_y-3}. \end{aligned} \tag{4.55}$$

The second inequality of (4.55) is valid since $i+1-k > 0$ and $i-m+1-k > 0$, so the terms in parentheses in the second member with exponents $d_y - 1 < 0$ are both smaller than 1. Similarly, j can be replaced by $j-p$ without diminishing the sum since its exponent is negative. Inserting the bound in (4.55) in A_{11n} in place of B_{11} and bounding the sums by integral approximation gives

$$\begin{aligned} n^2K(n)^2A_{11n} &\ll \sum_{i=1}^{n-1}\sum_{k=1}^{i}\sum_{m=0}^{i-1}\sum_{p=0}^{k-1}(k-p)^{2d_x+2d_y-3} \\ &\ll \sum_{i=1}^{n-1} i\sum_{k=1}^{i} k^{2d_x+2d_y-2} \\ &\ll \sum_{i=1}^{n-1} i^{2d_x+2d_y} = O(n^{2d_x+2d_y+1}). \end{aligned} \tag{4.56}$$

In B_{12} on the other hand, $j+1-k > 0$ and $i > m \geq 0$ and so, similarly,

$$\begin{aligned} B_{12} &\ll \sum_{j=k}^{\infty} j^{d_x-1}(j-p)^{d_x-1} i^{d_y-1}\left(1+\frac{1-k+j}{i}\right)^{d_y-1} \\ &\qquad\qquad \times (i-m)^{d_y-1}\left(1+\frac{1-k+j}{i-m}\right)^{d_y-1} \\ &\ll (k-p)^{2d_x-1} i^{d_y-1}(i-m)^{d_y-1} \end{aligned} \tag{4.57}$$

where in this case the terms in large parentheses with negative exponents are vanishing as $j \to \infty$. Substituting the bounding terms of (4.57) into A_{12n} gives

$$n^2K(n)^2A_{12n} \ll \sum_{i=1}^{n-1}\sum_{k=1}^{i} i^{d_y-1}\sum_{m=0}^{i-1}(i-m)^{d_y-1}\sum_{p=0}^{k-1}(k-p)^{2d_x-1}$$

$$\ll \sum_{i=1}^{n-1} i^{d_y-1}\sum_{k=1}^{i} i^{d_y}k^{2d_x} = O(n^{2d_x+2d_y+1}). \tag{4.58}$$

According to (4.52) it follows that $A_{1n} = O(n^{-1})$.

Next, consider A_{2n} for those cases with $i > k$. First,

$$B_{21} \ll \sum_{j=0}^{i-m-k} j^{d_x+2d_y-3}(j-p)^{d_x-1}\left(1+\frac{i+1-k}{j}\right)^{d_y-1}$$

$$\times\left(1+\frac{i-m+1-k}{j}\right)^{d_y-1}$$

$$\ll (i-m-k)^{2d_x+2d_y-3}. \tag{4.59}$$

The terms in large parentheses with negative exponents don't exceed 1, as in B_{11}, and in this case $p < 0$ so replacing $j-p$ by j does not decrease the bound. Thus,

$$n^2K(n)^2A_{21n} \ll \sum_{i=1}^{n-1}\sum_{k=1}^{i}\sum_{m=0}^{i-k-1}\sum_{p=k-i+m}^{-1}(i-m-k)^{2d_x+2d_y-3}$$

$$\ll \sum_{i=1}^{n-1}\sum_{k=1}^{i}\sum_{m=0}^{i-k-1}(i-m-k)^{2d_x+2d_y-2}$$

$$\ll \sum_{i=1}^{n-1}\sum_{k=1}^{i}(i-k)^{2d_x+2d_y-1} = O(n^{2d_x+2d_y+1}). \tag{4.60}$$

By similar reasoning,

$$B_{22} \ll \sum_{j=i-m-k+1}^{\infty} j^{d_x-1}(j-p)^{d_x-1}i^{d_y-1}\left(1+\frac{1-k+j}{i}\right)^{d_y-1}$$

$$\times(i-m)^{d_y-1}\left(1+\frac{1-k+j}{i-m}\right)^{d_y-1}$$

$$\ll (i-m-k)^{2d_x-1}i^{d_y-1}(i-m)^{d_y-1} \tag{4.61}$$

and so

$$n^2K(n)^2A_{22n} \ll \sum_{i=1}^{n-1} i^{d_y-1} \sum_{k=1}^{i} \sum_{m=0}^{i-k-1} (i-m)^{d_y-1} \sum_{p=k-i+m}^{-1} (i-m-k)^{2d_x-1}$$

$$\ll \sum_{i=1}^{n-1} i^{d_y-1} \sum_{k=1}^{i} \sum_{m=0}^{i-k-1} (i-m-k)^{2d_x+d_y-1}$$

$$\ll \sum_{i=1}^{n-1} i^{d_y-1} \sum_{k=1}^{i} (i-k)^{2d_x+d_y} = O(n^{2d_x+2d_y+1}). \tag{4.62}$$

The second inequality of (4.62) is valid since the bound is not decreased by replacing $(i-m)^{d_y-1}$ by $(i-m-k)^{d_y-1}$. In view of (4.60) and (4.62), $A_{2n} = O(n^{-1})$ according to (4.54). Hence by (4.50) the proof is complete. ∎

Chapter 5
Stochastic Integrals

5.1 Mean Deviations

The next objective is to study the asymptotic behaviour of the terms G_{1n} in (4.13) and G_{3n} in (4.15). The limiting forms of these terms will be denoted by Ξ_1 and Ξ_3 respectively and their sum by Ξ. The task of the present chapter is to determine the forms of the random variables Ξ_1 and Ξ_3, while Chapter 6 tackles the main business of proving that $G_n - \mathrm{E}(G_n) \to_{\mathrm{d}} \Xi$, subject to Assumption **4.1** and some additional conditions which include $d_x + d_y > -\frac{1}{2}$.

The first step is to express G_{1n} and G_{3n} by some judicious rearrangement of terms as $\mathcal{F}_n$-adapted stochastic processes. For the case of G_{1n} consider the expression in (4.13). For a given k and i, imagine the terms of this sum as entries in a rectangular table with infinitely many rows, with j the row index and l the column index. The first row has $i - k + 1$ entries, the second row has $i - k + 2$ entries, and so forth. In (4.13) the inner sum is of row elements across columns. Interchanging the order of summation over j and l, so that now the inner sum is of column elements by rows, gives

$$G_{1n} = \frac{1}{nK(n)} \sum_{i=1}^{n-1} \sum_{k=1}^{i} \sum_{l=0}^{\infty} \sum_{j=\max\{0,l+k-i\}}^{\infty} b_j c_l u_{k-j} w_{i+1-l}. \tag{5.1}$$

Next, let $p = k - j$ and $m = i - l$. Making these substitutions to eliminate j and l, note that (5.1) is the same as

$$G_{1n} = \frac{1}{nK(n)} \sum_{i=1}^{n-1} \sum_{m=-\infty}^{i} c_{i-m} w_{m+1} \sum_{k=1}^{i} \sum_{p=-\infty}^{\min\{k,m\}} b_{k-p} u_p. \tag{5.2}$$

The final step is again to interchange orders of summation, this time over m and i, and also over k and p. The result of this rearrangement is

Asymptotics for Fractional Processes. James Davidson, Oxford University Press. © James Davidson (2025).
DOI: 10.1093/9780198955207.003.0005

$$G_{1n} = \frac{1}{nK(n)} \sum_{m=-\infty}^{n-1} w_{m+1} \left(\sum_{p=-\infty}^{m} u_p \sum_{i=\max\{1,m\}}^{n-1} c_{i-m} \sum_{k=\max\{1,p\}}^{i} b_{k-p} \right)$$

$$= \frac{1}{\sqrt{n}} \sum_{m=-\infty}^{n-1} q_{nm} w_{m+1} \tag{5.3}$$

where

$$q_{nm} = \frac{1}{\sqrt{n}K(n)} \sum_{p=-\infty}^{m} a_{nmp} u_p \tag{5.4}$$

and, restoring the original subscripts $l = i - m$ and $j = k - p$,

$$a_{nmp} = \sum_{l=\max\{0,1-m\}}^{n-1-m} c_l \left(\sum_{j=\max\{0,1-p\}}^{l+m-p} b_j \right). \tag{5.5}$$

Take care to distinguish this usage of the symbol a from that in (2.26), noting the three subscripts in place of the earlier two.

G_{3n} is rearranged in much the same manner. Defining $p = k - j$ and $m = i + 1 - l$ (noting that now $m < p < n$) and identifying these subscripts with the variables w_m and u_p gives

$$G_{3n} = \frac{1}{nK(n)} \sum_{i=1}^{n-1} \sum_{k=1}^{i} \sum_{p=-\infty}^{k} \sum_{m=-\infty}^{p-1} b_{k-p} c_{i+1-m} u_p w_m.$$

Next, interchanging orders of summation, first over i and k and then over k and p, has the result

$$G_{3n} = \frac{1}{nK(n)} \sum_{p=-\infty}^{n-1} u_p \left(\sum_{m=-\infty}^{p-1} w_m \sum_{k=\max\{p,1\}}^{n-1} b_{k-p} \sum_{i=k}^{n-1} c_{i+1-m} \right)$$

$$= \frac{1}{\sqrt{n}} \sum_{p=-\infty}^{n-1} h_{np} u_p \tag{5.6}$$

where

$$h_{np} = \frac{1}{\sqrt{n}K(n)} \sum_{m=-\infty}^{p-1} e_{npm} w_m \tag{5.7}$$

and (restoring original subscripts)

$$e_{npm} = \sum_{j=\max\{0,1-p\}}^{n-1-p} b_j \left(\sum_{l=j+p+1-m}^{n-m} c_l \right). \tag{5.8}$$

Equations (5.3) and (5.4), and (5.6) and (5.7), show G_{1n} and G_{3n} to have the appearance of sample covariances of a relatively familiar type, apart from the infinite order of the sum. Specifically, these are the covariances between independent processes, w_i and u_i respectively, and moving average processes lagged one period. That their means are zero has been arranged by construction, with the contemporaneous pairs removed to G_{2n}. What is unusual is the construction of the moving average weights. These are in general non-summable and hence affiliated with unit root processes, although here the weights a_{nmp} in (5.4) and e_{npm} in (5.7) are dependent both on their positions in the sequence and on the terminal dates, rather than being just unity or zero.

To get a feel for the formula for G_{1n}, as the exemplar case, it is of interest to see what happens to (5.5) in the respective special cases $d_x = 0$, so that specifically $b_j = 0$ for $j > 0$, and $d_y = 0$ so that $c_l = 0$ for $l > 0$. In the first case, noting that $p \leq m$, $a_{nmp} = 1 + c_1 + \cdots + c_{n-1-m}$ if $p > 0$ and $a_{nmp} = 0$ otherwise. In the second case, $a_{nmp} = b_{\max\{0,1-p\}} + \cdots + b_{m-p}$ if $m > 0$, which is the same as $a_{np}(m/n, 0)$ from (2.26), and $a_{nmp} = 0$ otherwise. In the two cases combined, noting $b_0 = 1$, $a_{nmp} = 1$ for $p = 1, \ldots, m$ and $1 \leq m \leq n - 1$, and $a_{nmp} = 0$ otherwise so q_{nm} reduces, as expected, to a random walk initialized at $p = 1$.

5.2 Integral Approximations

The next step is to evaluate the limiting forms of the arrays $K(n)^{-1} a_{nmp}$ and $K(n)^{-1} e_{npm}$ as $n \to \infty$. The following theorem is helpful for connecting the limiting behaviour of the partial sums appearing in (5.5) and (5.8) with the assumptions about the sequences $\{b_j\}$ and $\{c_l\}$.

5.1 Theorem If $b > a \geq 0$, $\alpha > 0$ and $L(n)$ is slowly varying at ∞,

$$\frac{1}{n^{\alpha} L(n)} \sum_{j=[na]}^{[nb]-1} j^{\alpha-1} L(j) = \frac{b^{\alpha} - a^{\alpha}}{\alpha} + o(1) \tag{5.9}$$

as $n \to \infty$.

Proof $L(j)/L(n) = 1 + o(1)$ for $[na] \leq j < [nb]$ and $a > 0$. In the case $a = 0$ there are terms in the sum for which j is finite in the limit and so $L(j)/L(n) \nrightarrow 1$ as $n \to \infty$. However, since the sum in (5.9) diverges these terms are of small order in n and omitting them does not affect the limit. It is therefore valid to conclude that for $a \geq 0$,

$$\frac{1}{n^\alpha L(n)}\sum_{j=[na]}^{[nb]-1} j^{\alpha-1}L(j) = \frac{1}{n}\sum_{j=[na]}^{[nb]-1}\left(\frac{j}{n}\right)^{\alpha-1} + o(1).$$

Taylor's expansions of the function $(\cdot)^\alpha/\alpha$ around j/n specify the relations

$$\frac{1}{\alpha}\left(\frac{j+1}{n}\right)^\alpha = \frac{1}{\alpha}\left(\frac{j}{n}\right)^\alpha + \frac{1}{n}\left(\frac{j}{n}\right)^{\alpha-1} + O(n^{-2}).$$

The proof is completed in view of the telescoping sum,

$$\sum_{j=[na]}^{[nb]-1}\frac{1}{\alpha}\left(\left(\frac{j+1}{n}\right)^\alpha - \left(\frac{j}{n}\right)^\alpha\right) = \frac{1}{\alpha}\left(\left(\frac{[nb]}{n}\right)^\alpha - \left(\frac{[na]}{n}\right)^\alpha\right). \quad ∎$$

A fundamental assumption to be carried through the development in this section is that $0 \le d_y < \frac{1}{2}$ in the discussion of G_{1n}, and also that $0 \le d_x < \frac{1}{2}$ in the treatment of G_{3n}. This is because antipersistence calls for a different treatment of the limit formulae. It is easiest from the point of view of exposition to deal with these two analyses separately and the required variations are to be found in §5.4.

Consider first the case of G_{1n}. Take care to note that the pair of symbols s and t, which in Chapters 2 and 3 played the roles of the upper and lower bounds of an interval so that $s > t$, are here re-used in a different context. In this case $s \le t$, although they will again change places in the analysis of G_{3n}.

5.2 Lemma If $|d_x| < \frac{1}{2}$ and $d_x+d_y > 0$, $K(n)^{-1}a_{n[nt][ns]} = A(t,s)+o(1)$ as $n \to \infty$ for real-valued indices t, s with $-\infty < s \le t < 1$, where

(i) if $0 < d_y < \frac{1}{2}$,

$$A(t,s) = d_y\int_{\max\{0,-t\}}^{1-t} v^{d_y-1}(v+t-s)^{d_x}\mathrm{d}v$$
$$- 1_{\{s<0\}}(-s)^{d_x}\left((1-t)^{d_y} - 1_{\{t<0\}}(-t)^{d_y}\right). \tag{5.10}$$

(ii) if $d_y = 0$,

$$A(t,s) = 1_{\{t\ge0\}}\left((t-s)^{d_x} - 1_{\{s<0\}}(-s)^{d_x}\right). \tag{5.11}$$

If $d_x + d_y \le 0$, the same conclusions hold only for $-\infty < s < t < 1$.

Proof Considering the components of equation (5.5), define u and v by $j = [nu]$ and $l = [nv]$. For part (i), if $d_x > 0$ then, recalling $b_j \sim d_x j^{d_x-1} L_x(j)$ as in (4.1), write

$$\frac{b_{[nu]}}{n^{d_x} L_x(n)} \sim d_x \frac{u^{d_x-1}}{n} \tag{5.12}$$

and by Theorem **5.1**,

$$\frac{1}{n^{d_x} L_x(n)} \sum_{j=\max\{0,1-[ns]\}}^{[nv]+[nt]-[ns]} b_j \sim (v+t-s)^{d_x} - 1_{\{s<0\}}(-s)^{d_x}. \tag{5.13}$$

If $d_x < 0$, then in view of (1.20) it is possible to write

$$\frac{b_{[nu]}}{n^{d_x} L_x(n)} \sim (u+1/n)^{d_x} - u^{d_x} \tag{5.14}$$

and in this case equality (5.13) follows directly. In the case $d_x = 0$ the b_j are at most summable giving

$$\sum_{j=\max\{0,1-[ns]\}}^{[nv]+[nt]-[ns]} b_j = \begin{cases} O(1), & s \geq 0 \\ o(1), & s < 0. \end{cases}$$

The sum could be assigned the limiting value $1 - 1_{\{s<0\}}$ by choice of normalization.

Similarly,

$$\frac{c_{[nv]}}{n^{d_y} L_y(n)} \sim d_y \frac{v^{d_y-1}}{n}. \tag{5.15}$$

Therefore, (5.5) implies $K(n)^{-1} a_{n[nt][ns]} = A(t,s) + o(1)$ with

$$A(t,s) = d_y \int_{\max\{0,-t\}}^{1-t} v^{d_y-1} \left((v+t-s)^{d_x} - 1_{\{s<0\}}(-s)^{d_x} \right) \mathrm{d}v \tag{5.16}$$

which matches the formula in (5.10), provided the integral in question exists.

There are three cases to consider. If $d_x > 0$ then $v^{d_y-1}(v+t-s)^{d_x} \leq v^{d_y-1}(1-s)^{d_x}$ since $v \leq 1-t$ and v^{d_y-1} is integrable by assumption. If $d_x < 0$ but at the same time $d_x + d_y > 0$, the integrand is bounded above at the point $t = s$ and $v^{d_x+d_y-1}$ is integrable. In these cases the integral exists for all $-\infty < s \leq t < 1$. If $d_x + d_y \leq 0$, in which case $d_x < 0$, the integral exists for $-\infty < s < t < 1$, but there is a singularity at the point $s = t$. There, the integral in (5.10) becomes

$$d_y \int_{\max\{0,-t\}}^{1-t} v^{d_x+d_y-1} dv$$

and diverges at 0.

For part (ii), if $c_l = 0$ for $l > 0$ or $\{c_l\}$ is a summable sequence, $K(n)^{-1} a_{n[nt][ns]}$ has the form of (5.13) with $v = 0$, provided also that $t \geq 0$ so that c_0 appears in the sum. If $t = s$ the singularity in (5.11) when $d_x < 0$ is evident. ∎

The point $t = 1$ has been excluded from this result but it is evident that $A(1, s)$ vanishes unless $d_y = 0$, in which case it is given by (5.11).

To help to interpret the singularity write $m = [nt]$ and $p = [ns]$, and consider how a_{nmp} behaves in the case $d_x + d_y < 0$, which here means $d_x < 0$. If $p < m$, or in other words $[nt] - [ns] = O(n)$, the sum of the b_j has a minimum of $[nt] - [ns]$ terms and converges to zero as $n \to \infty$. In this case, normalization by $n^{-d_x-d_y}$ has the desired result. However, if $s = t$ and hence $p = m$, then according to (5.5) the first few terms of the sum have the form $c_l(b_0 + \cdots + b_l)$ for $l = 0, 1, 2, \ldots$. These initial coordinates do not vanish as n increases and so are ill-adapted to the normalization by $K(n)$. Dividing by $n^{d_x+d_y}$ must result in divergence.

This phenomenon links directly to the differing orders of magnitude of the mean and the mean deviation processes examined in Theorem **4.7**, and also connects with a version of the boundedness in probability issue that arose in Theorem **3.5**. This is to be examined in the next chapter in Theorem **6.13** and gives rise to a moment restriction comparable with Assumption **3.1**.

Referring to equation (5.3), an important fact is that when $p = m$, u_p is separated by only one time period from w_{m+1}. Let the substitutions $p = k - j$ and $m = i - l$ be reversed, and then refer back to the threefold decomposition of G_n. $p = m$ is equivalent to $l = i + j - k$ which defines the boundary between the sums G_{1n} in (4.13) and G_{2n} in (4.14). The simplest solution to the singularity is to impose the restriction $s < t$ on the integral, which amounts to relocating problematic terms from G_{1n} to G_{2n} where the normalization is appropriate according to §4.4. Equations (5.12) and (5.15) show that after normalization the contributions of individual terms to the sums are $O(1/n)$. A collection numbering at most $o(n)$ is of small order in the limit, so the effect of the transfer on the limiting expectation can be neglected.

The following substitutions will prove helpful in the sequel. Define the functions

$$Z_1^A(t, s) = d_y \int_0^1 \tau^{d_y-1} \left(1 - \frac{t-1}{t-s}\tau\right)^{d_x} d\tau \qquad (5.17)$$

for $-\infty < t < 1$ and $-\infty < s < t$, and

$$Z_2^A(t,s) = d_y \int_0^1 \tau^{d_y-1}\left(1 - \frac{t}{t-s}\tau\right)^{d_x} \mathrm{d}\tau \tag{5.18}$$

for $-\infty < t < 0$ and $-\infty < s < t$. These are the integral forms of the hypergeometric functions $F(a, b; c; z)$ defined in (B.24), where $a = -d_x$, $b = d_y$, and $c = d_y + 1$, with $z = (t-1)/(t-s)$ in the case of Z_1^A and $z = t/(t-s)$ in the case of Z_2^A. For $|z| \leq 1$ these functions are represented by the Gauss hypergeometric series (B.22) and are defined by analytic continuation elsewhere. The integrals exist with real arguments if $b > 0$ and $c > b$ and since $z < 0$ in both cases, with $d_y > 0$ the integrals are well defined. Chapter 15 of [1] has additional details.

With these definitions, a convenient way to visualize the behaviour of the function (5.10) as it depends on s and t is to substitute (5.17) and (5.18) into the formula making the changes of variable $\tau = v/(1-t)$ in the first term and $\tau = v/(-t)$ in the second term. In this way, write the integral in (5.10) as

$$\begin{aligned} d_y &\int_{\max\{0,-t\}}^{1-t} v^{d_y-1}(v+t-s)^{d_x}\mathrm{d}v \\ &= (t-s)^{d_x}\left((1-t)^{d_y}Z_1^A(t,s) - 1_{\{t<0\}}(-t)^{d_y}Z_2^A(t,s)\right). \end{aligned} \tag{5.19}$$

The function in (5.19) is well defined everywhere except at the singular point. Note in particular that when $d_x < 0$, $Z_1^A(t,s)$ and $Z_2^A(t,s)$ tend to zero as $s \to t$. This representation will be put to use in proving the functional central limit theorem.

Next, consider the case of G_{3n}. The following result closely parallels Lemma **5.2** except that d_x and d_y switch roles and $s \geq t$, reflecting the switch of roles of p and m.

5.3 Lemma If $|d_y| < \frac{1}{2}$ and $d_x + d_y > 0$, $K(n)^{-1}e_{n[ns][nt]} = E(s,t) + o(1)$ as $n \to \infty$ for real-valued indices t, s with $-\infty < t \leq s < 1$, where

(i) if $0 < d_x < \frac{1}{2}$,

$$\begin{aligned} E(s,t) = (1-t)^{d_y}&\left((1-s)^{d_x} - 1_{\{s<0\}}(-s)^{d_x}\right) \\ &- d_x \int_{\max\{0,-s\}}^{1-s} v^{d_x-1}(v+s-t)^{d_y}\mathrm{d}v. \end{aligned} \tag{5.20}$$

(ii) if $d_x = 0$,

$$E(s,t) = 1_{\{s\geq 0\}}\left((1-t)^{d_y} - (s-t)^{d_y}\right). \tag{5.21}$$

If $d_x + d_y \leq 0$, the same conclusions hold only for $-\infty < t < s < 1$.

Proof For part (i), define v by $j = [nv]$. Applying Theorem **5.1** to (5.8), similarly to (5.13),

$$\frac{1}{n^{d_y} L_y(n)} \sum_{l=[nv]+[ns]+1-[nt]}^{n-[nt]} c_l \sim (1-t)^{d_y} - (v+s-t)^{d_y}$$

so that

$$E(s,t) = d_x \int_{\max\{0,-s\}}^{1-s} v^{d_x-1}\left((1-t)^{d_y} - (v+s-t)^{d_y}\right) \mathrm{d}v.$$

After rearrangement this formula matches (5.20). The integral exists with $d_x + d_y > 0$ by reasoning similar to **5.2**, but if $d_x + d_y \leq 0$ so that necessarily $d_y < 0$, there is again a singularity at $s = t$.

For part (ii) the formula is found directly from (5.8) and the singularity is evident from (5.21). ∎

Analogously to (5.17) and (5.18), the hypergeometric integrals for this case are

$$Z_1^E(s,t) = d_x \int_0^1 \tau^{d_x-1}\left(1 - \frac{s-1}{s-t}\tau\right)^{d_y} \mathrm{d}\tau$$

for $-\infty < s < 1$ and $-\infty < t < s$, and

$$Z_2^E(s,t) = d_x \int_0^1 \tau^{d_x-1}\left(1 - \frac{s}{s-t}\tau\right)^{d_y} \mathrm{d}\tau$$

for $-\infty < s < 0$ and $-\infty < t < s$. These are well defined when $d_x > 0$ and similarly to (5.19), substitutions with respective changes of variable $\tau = v/(1-s)$ and $\tau = v/(-s)$ yield the form

$$d_x \int_{\max\{0,-s\}}^{1-s} v^{d_x-1}(v+s-t)^{d_y} \mathrm{d}v$$
$$= (s-t)^{d_y}\left((1-s)^{d_x} Z_1^E(s,t) - 1_{\{s<0\}}(-s)^{d_x} Z_2^E(s,t)\right). \tag{5.22}$$

5.3 Heuristic Representation

Let U and W denote Brownian motions on the real line segment $(-\infty, 1]$, setting $U(0) = W(0) = 0$ (which involves no loss of generality) and with variances $\mathrm{E}(U(1)^2) = \sigma_u^2$ and $\mathrm{E}(W(1)^2) = \sigma_w^2$ and covariance $\mathrm{E}(U(1)W(1)) = \sigma_{uw}$.

The substitutions $\mathrm{d}U(s)$ for $u_{[ns]}/\sqrt{n}$ and $\mathrm{d}W(t)$ for $w_{[nt]}/\sqrt{n}$ can then be made to develop an asymptotic approximation to (5.3). Writing $\mathcal{F}_n(t) = \mathcal{F}_{[nt]} \in \boldsymbol{F}$, as defined following Assumption **4.1**, let $\{\mathcal{F}(t), t \in \mathbb{R}\}$ denote the filtration to which U and W are adapted, where $\mathcal{F}(t)$ is the limiting case of $\mathcal{F}_n(t)$ as $n \to \infty$.

The limit of the normalized random variable G_{1n} in (5.3) can be expressed heuristically as

$$\Xi_1 = \int_{-\infty}^{1} Q(t)\mathrm{d}W(t) \tag{5.23}$$

where, with $A(t,s)$ defined in (5.10) or (5.11),

$$Q(t) = \int_{-\infty}^{t} A(t,s)\mathrm{d}U(s) \tag{5.24}$$

is a $\mathcal{F}(t)$-adapted stochastic process in continuous time. Note the remark on page 20 concerning the interpretation of infinite-lag processes, which also applies to (5.24) and (5.23). Also bear in mind the resolution of the singularity under antipersistence by formally excluding the points $s = t$ from (5.24). What this means in practice is that different normalizations arise when these boundary points enter calculations, as will become clear in the proof of Lemma **6.12** and subsequently.

For future reference, useful implications of formula (5.24) include the variance of $Q(t)$ for $-\infty < t \le 1$,

$$\mathrm{E}(Q(t)^2) = \sigma_u^2 \int_{-\infty}^{t} A^2(t,s)\mathrm{d}s \tag{5.25}$$

and also the variance of an increment,

$$\begin{aligned} &\mathrm{E}(Q(t+\delta) - Q(t))^2 \\ &= \sigma_u^2 \int_{t}^{t+\delta} A^2(t+\delta,s)\mathrm{d}s + \sigma_u^2 \int_{-\infty}^{t} (A(t+\delta,s) - A(t,s))^2\,\mathrm{d}s. \end{aligned} \tag{5.26}$$

In the case $d_y = 0$, according to Lemma **5.2**(ii) a_{nmp} specializes to $a_{np}(m/n,0)$ from (2.26) and (5.4) becomes

$$q_{nm} = \frac{1}{n^{d_x+1/2}L_x(n)} \sum_{k=1}^{m} x_k \sim X_n(m/n) \tag{5.27}$$

from (2.28). In other words, $Q(t) = X(t)$ for $t \geq 0$ and $Q(t) = 0$ for $t < 0$, where X is fBm as defined by (2.1) with $d = d_x$. In this case

$$\Xi_1 = \int_0^1 X(t)\mathrm{d}W(t) \tag{5.28}$$

which is the Itô integral with respect to W of a fBm integrand.

In the same way, the limit of G_{3n} can be expressed in the form

$$\Xi_3 = \int_{-\infty}^1 H(s)\mathrm{d}U(s) \tag{5.29}$$

where the $\mathcal{F}(s)$-adapted integrand process is

$$H(s) = \int_{-\infty}^s E(s,t)\mathrm{d}W(t). \tag{5.30}$$

When $d_y = 0$, $E(s,t) = 0$ for all s and t. The variable Ξ_3 arises only in the case of a fractional integrator function, otherwise $\Xi = \Xi_1$ and this term has the form shown in (5.28).

The important fact is that both Ξ_1 and Ξ_3 (where it exists) are stochastic integrals of $\mathcal{F}$-adapted integrand processes with respect to $\mathcal{F}$-adapted Brownian motions. These integrals are of the Itô type. Subject to sufficient regularity conditions on the integrands, essentially those of finite variances and a.s. continuity, plus the validity of mean-squared approximations by integrals with finite domain of integration, they may be analyzed in the fashion familiar from the unit root analysis. The important difference is that, as is apparent from (5.26), the integrand processes Q and H are not Brownian motions and have dependent increments.

5.4 Antipersistent Integrators

When $d_y < 0$, while the formula in (5.5) is valid the integral approximation in (5.10) fails and another way of constructing an asymptotic approximation must be found. In the case $d_x < 0$, the same is true of (5.8) and (5.20).

Focusing first on G_{1n}, an alternative form of expression (5.5) exists whenever it is possible to write $c_l = c_l^* - c_{l-1}^*$ for $l > 0$ for a suitably defined sequence $\{c_l^*\}$, with $c_0^* = c_0 = 1$. Then, if $c_l \sim d_y l^{d_y-1}L_y(l)$ for $l > 0$ with $-\frac{1}{2} < d_y < 0$ it must be the case that $c_l^* \sim l^{d_y}L_y(l)$, as explained in the discussion leading to

(1.20). With this substitution the double sum of (5.5) can be written as

$$\begin{aligned}a_{nmp} &= 1_{\{m>0\}}c_0^* \sum_{j=\max\{0,1-p\}}^{m-p} b_j + \sum_{l=\max\{1,1-m\}}^{n-1-m} (c_l^* - c_{l-1}^*) \sum_{j=\max\{0,1-p\}}^{l+m-p} b_j \\ &= c_{n-1-m}^* \sum_{j=\max\{0,1-p\}}^{n-1-p} b_j - \sum_{l=\max\{0,-m\}}^{n-2-m} c_l^* b_{l+1+m-p} \qquad (5.31)\end{aligned}$$

where the second equality is obtained by cancelling the pairs of matching terms with opposite signs.

A simple application of the formula in (5.31), for comparison with (5.27), is the case $d_y = 0$. Specifically, if $c_l = 0$ for $l > 0$ then $c_l^* = 1$ for all $l > 0$. Formula (5.31) therefore yields $a_{nmp} = b_{\max\{0,1-p\}} + \cdots + b_{m-p}$ for $1 \le m \le n-1$ and 0 otherwise, which matches the corresponding case of (5.5).

While (5.31) is identical to (5.5), the same set of terms simply being ordered in a different way, the asymptotic approximation needs to be constructed differently. Therefore, although the notation a_{nmp} applies in each case, the expression $A(t,s)$ in (5.10) has to be replaced by a different one, to be denoted $A^*(t,s)$. The same integral approximation arguments apply.

5.4 Lemma If $|d_x| < \frac{1}{2}$ and $-\frac{1}{2} < d_y < 0$, $K(n)^{-1}a_{n[nt][ns]} = A^*(t,s) + o(1)$ as $n \to \infty$ where for real-valued indices t, s,

$$\begin{aligned}A^*(t,s) = (1-t)^{d_y}\left((1-s)^{d_x} - 1_{\{s<0\}}(-s)^{d_x}\right) \\ - d_x \int_{\max\{0,-t\}}^{1-t} v^{d_y}(v+t-s)^{d_x-1} dv \qquad (5.32)\end{aligned}$$

with $-\infty < s \le t < 1$ if $d_x + d_y > 0$. If $d_x + d_y \le 0$, the same conclusion holds only for $-\infty < s < t < 1$.

Proof The first term of (5.32) is easily constructed from that of (5.31) since

$$\frac{c_{n-1-[nt]}^*}{n^{d_y} L_y(n)} \sim (1-t)^{d_y}.$$

The second term is found by the argument analogous to that of Lemma **5.2** since

$$\frac{c_{[nv]}^*}{n^{d_y} L_y(n)} \sim v^{d_y}.$$

Integrability holds in this case with $d_x + d_y > 0$ because with $d_x - 1 < 0$ the integrand is bounded above by $v^{d_y+d_x-1}$. In the case $d_x + d_y \leq 0$, the singularity at $s = t$ is noted as before. ∎

The hypergeometric functions appropriate to this case are

$$Z_1^{A*}(t,s) = (1 + d_y) \int_0^1 \tau^{d_y} \left(1 - \frac{t-1}{t-s}\tau\right)^{d_x-1} d\tau \tag{5.33}$$

for $-\infty < t < 1$ and $-\infty < s < t$, and

$$Z_2^{A*}(t,s) = (1 + d_y) \int_0^1 \tau^{d_y} \left(1 - \frac{t}{t-s}\tau\right)^{d_x-1} d\tau \tag{5.34}$$

for $-\infty < t < 0$ and $-\infty < s < t$. In these cases of $F(a, b; c; z)$, $b = 1 + d_y > 0$ and $c = 2 + d_y$. With changes of variable $\tau = v/(1 - t)$ in (5.33) and $\tau = v/(-t)$ in (5.34), the integral representation

$$\begin{aligned} d_x \int_{\max\{0,-t\}}^{1-t} & v^{d_y}(v + t - s)^{d_x-1} dv \\ &= \frac{d_x}{1 + d_y}(t - s)^{d_x-1}\left((1 - t)^{d_y+1} Z_1^{A*}(t,s) - 1_{\{t<0\}}(-t)^{d_y+1} Z_2^{A*}(t,s)\right) \end{aligned} \tag{5.35}$$

is well defined for $-\infty < s < t < 1$.

Moving on to the case of G_{3n}, the modification corresponding to (5.31) can be performed on the formula in (5.8) for the case $d_x < 0$. Define the sequence b_j^* by $b_0^* = 1$ and $b_j = b_j^* - b_{j-1}^*$ for $j > 0$, with $b_j^* \sim j^{d_x} L_x(j)$. Noting $p > m$ in this case, the double sum in (5.8) becomes

$$\begin{aligned} e_{nmp} &= 1_{\{p>0\}} b_0^* \sum_{l=p+1-m}^{n-m} c_l + \sum_{j=\max\{1,1-p\}}^{n-1-p} (b_j^* - b_{j-1}^*) \sum_{l=j+p+1-m}^{n-m} c_l \\ &= \sum_{j=\max\{0,1-p\}}^{n-1-p} b_j^* c_{j+p+1-m} - 1_{\{p\leq 0\}} b_{-p}^* \sum_{l=2-m}^{n-m} c_l \end{aligned} \tag{5.36}$$

where the second equality is obtained by cancelling equal and oppositely signed terms. Arguments closely paralleling Lemma **5.4** (not given explicitly in this case) yield the following.

5.5 Lemma If $|d_y| < \frac{1}{2}$ and $-\frac{1}{2} < d_x < 0$, $K(n)^{-1}e_{n[nt][ns]} = E^*(s,t) + o(1)$ as $n \to \infty$ where for real-valued indices t, s,

$$E^*(s,t) = d_y \int_{\max\{0,-s\}}^{1-s} v^{d_x}(v+s-t)^{d_y-1}\mathrm{d}v \\ - 1_{\{s\leq 0\}}(-s)^{d_x}\left((1-t)^{d_y} - (-t)^{d_y}\right) \tag{5.37}$$

with $-\infty < t \leq s < 1$ if $d_x + d_y > 0$. If $d_x + d_y \leq 0$, the same conclusion holds only for $-\infty < t < s < 1$. □

In the approximation arguments to appear in Chapter 6, the expressions A^* and E^* are used in place of A and E to deal with antipersistent integrators in Ξ_1 and Ξ_3 respectively. With this exception, the arguments are very similar in the two cases.

5.5 Integration by Parts

At this point, recall from §4.2 the suggested notation $G_n^{xy} = G_{1n}^{xy} + G_{2n}^{xy} + G_{3n}^{xy}$ in place of G_n, when the distinction needs to be made. In this framework, when $d_x + d_y > 0$ it is shown by Theorem **4.8** that $G_{2n}^{xy} \to_{L_2} \sigma_{uw}\lambda_{xy}$. It is thus far conjectured (and to be proved formally in Chapter 6) that G_{1n}^{xy} and G_{3n}^{xy} converge in distribution to limits Ξ_{1xy} and Ξ_{3xy} with sum Ξ_{xy} where these symbols now stand in for the Ξ_1, Ξ_3 and Ξ used previously. The complementary cases are defined as Ξ_{1yx}, Ξ_{3yx}, and Ξ_{yx} in which the roles of the variables x and y are interchanged. λ_{yx} has already been defined on page 76 and in a similar fashion, the notations A_{xy}, A_{yx} and A_{xy}^*, A_{yx}^*, and also E_{xy}, E_{yx} and E_{xy}^*, E_{yx}^* will be used to distinguish the integrals in (5.10), (5.32), (5.20), and (5.37) from their complementary cases.

With these changes in place, let X and Y denote the fBms having the form of (2.1) with respective parameters d_x and d_y and respective driving processes U and W, as defined in §5.3. With $d_x + d_y > 0$ it appears natural to equate the random variable $\Xi_{xy} + \sigma_{uw}\lambda_{xy}$ with the stochastic integral $\int_0^1 X\mathrm{d}Y$ and likewise $\Xi_{yx} + \sigma_{uw}\lambda_{yx}$ with $\int_0^1 Y\mathrm{d}X$. However, for the designation 'integral' here to be appropriate it should be the case that the integration by parts rule

$$\int_0^1 X\mathrm{d}Y + \int_0^1 Y\mathrm{d}X = X(1)Y(1) \tag{5.38}$$

is obeyed.

Matters are less clear cut when $d_x + d_y \le 0$ in view of the antipersistent behaviour noted in §4.4. In particular, λ_{xy} and λ_{yx} don't exist and it doesn't appear possible to write an equation such as (5.38). What is known from (4.21) is that $\mathrm{E}(X(1)Y(1)) = \sigma_{uw}\Upsilon_{xy}$, where $\Upsilon_{xy} = \lambda_{xy} + \lambda_{yx}$ whenever this relation is well defined. The unambiguous requirement for a valid formulation, that covers every case with $|d_x| < \frac{1}{2}$ and $|d_y| < \frac{1}{2}$, is that

$$\Xi_{xy} + \Xi_{yx} + \sigma_{uw}\Upsilon_{xy} = X(1)Y(1). \tag{5.39}$$

To explore these questions, some interesting facts about the complementary expressions can be brought to light. First, from (5.23) and (5.29),

$$\Xi_{xy} = \int_{-\infty}^1 \int_{-\infty}^t A_{xy}(t,s)\mathrm{d}U(s)\mathrm{d}W(t) + \int_{-\infty}^1 \int_{-\infty}^s E_{xy}(s,t)\mathrm{d}W(t)\mathrm{d}U(s).$$

Swapping the variables, the arguments are all completely symmetric and

$$\Xi_{yx} = \int_{-\infty}^1 \int_{-\infty}^s A_{yx}(s,t)\mathrm{d}W(t)\mathrm{d}U(s) + \int_{-\infty}^1 \int_{-\infty}^t E_{yx}(t,s)\mathrm{d}U(s)\mathrm{d}W(t).$$

Adding these terms together gives

$$\begin{aligned}\Xi_{xy} + \Xi_{yx} &= \int_{-\infty}^1 \int_{-\infty}^t (A_{xy}(t,s) + E_{yx}(t,s))\mathrm{d}U(s)\mathrm{d}W(t) \\ &\quad + \int_{-\infty}^1 \int_{-\infty}^s (A_{yx}(s,t) + E_{xy}(s,t))\mathrm{d}W(t)\mathrm{d}U(s).\end{aligned} \tag{5.40}$$

Examine (5.10) and (5.11) and also the versions of (5.20) and (5.21) that are obtained by swapping x and y and also t and s. The integrals are equal and oppositely signed and so cancel in the sum. After some rearrangement it is found that

$$\begin{aligned}&A_{xy}(t,s) + E_{yx}(t,s) \\ &\quad = \left((1-t)^{d_y} - 1_{\{t<0\}}(-t)^{d_y}\right)\left((1-s)^{d_x} - 1_{\{s<0\}}(-s)^{d_x}\right).\end{aligned} \tag{5.41}$$

Given the symmetry of the two cases, it is as easily seen that

$$\begin{aligned} A_{yx}(s,t) &+ E_{xy}(s,t) \\ &= \left((1-s)^{d_x} - 1_{\{s<0\}}(-s)^{d_x}\right)\left((1-t)^{d_y} - 1_{\{t<0\}}(-t)^{d_y}\right) \end{aligned} \tag{5.42}$$

so that expressions (5.41) and (5.42) are actually identical.

The same manipulations can be applied to the formulae developed in §5.4 for the antipersistent case. Let $A^*_{yx}(s,t)$ and $E^*_{yx}(t,s)$ be the cases complementary to $A^*_{xy}(t,s)$ and $E^*_{xy}(s,t)$ from (5.32) and (5.37), with d_y and d_x and also s and t interchanged. The integral terms cancel as before and it can be verified by direct inspection that

$$A^*_{xy}(t,s) + E^*_{yx}(t,s) = A_{xy}(t,s) + E_{yx}(t,s) \tag{5.43}$$

and also

$$A^*_{yx}(s,t) + E^*_{xy}(s,t) = A_{yx}(s,t) + E_{xy}(s,t). \tag{5.44}$$

These identities contribute to the proof of the next theorem confirming the equality in (5.39). This is not an asymptotic result, U and W being assumed to be regular Brownian motions and the integrals otherwise depending solely on the d_x and d_y parameters and σ_{uw}. The symbol σ_{uw} appears in the limit in Theorem **4.4**, because the limit is derived under Assumption **4.1** in which the finite-sample shocks are serially independent. If these are in fact autocorrelated, which is the case to be developed in Chapter 8 under Assumption **8.1**, then σ_{uw} should be replaced by ω_{uw} standing for the long-run covariance. In the present context, nothing depends on these assumptions since the constant is merely an attribute of the joint distribution of (U, W) and the switch is a formality.

5.6 Theorem If $|d_x| < \frac{1}{2}$ and $|d_y| < \frac{1}{2}$ then (5.39) holds.

Proof Assume initially that $d_y \geq 0$ and $d_x \geq 0$, these being the cases to which formulae (5.41), (5.42), and (5.40) are relevant.

It will be helpful to introduce a compact notation. For a function $F_Y : (-\infty, 1] \mapsto \mathbb{R}$ and fractional process X (as defined in (2.1), with parameter d_x) define $\int F_Y \delta X$ by the formula

$$\int F_Y \delta X = \int_{-\infty}^{1} F_Y(\tau)\left((1-\tau)^{d_x} - 1_{\{\tau<0\}}(-\tau)^{d_x}\right) dU(\tau). \tag{5.45}$$

In particular, note on putting $F_Y = 1$ that $\int \delta X = X(1)$ from (2.1). Similarly, for a function F_X and fractional process Y with parameter d_y define

$$\int F_X \delta Y = \int_{-\infty}^{1} F_X(t)\left((1-t)^{d_y} - 1_{\{t<0\}}(-t)^{d_y}\right) \mathrm{d}W(t). \tag{5.46}$$

such that $\int \delta Y = Y(1)$. Let the roles of F_Y and F_X be played initially by the respective processes $\widetilde{Y} : (-\infty, 1] \mapsto \mathbb{R}$ and $\widetilde{X} : (-\infty, 1] \mapsto \mathbb{R}$, where

$$\widetilde{Y}(\tau) = \int_{-\infty}^{\tau} \left((1-t)^{d_y} - 1_{\{t<0\}}(-t)^{d_y}\right) \mathrm{d}W(t)$$

and

$$\widetilde{X}(t) = \int_{-\infty}^{t} \left((1-\tau)^{d_x} - 1_{\{\tau<0\}}(-\tau)^{d_x}\right) \mathrm{d}U(\tau).$$

Thus, note that $\widetilde{Y}(\tau) = Y(\tau)$ for $\tau \geq 0$ and $\widetilde{X}(t) = X(t)$ for $t \geq 0$, the difference being that these processes are defined over the entire domain indicated. It can be verified that $\int \widetilde{Y}\delta X$ and $\int \widetilde{X}\delta Y$ are Itô integrals with respective Brownian integrator processes U and W, with the important implication that they have zero means. Further, substituting the matching formulae from (5.41) and (5.42) into (5.40) yields the representation

$$\Xi_{xy} + \Xi_{yx} = \int \widetilde{X}\delta Y + \int \widetilde{Y}\delta X. \tag{5.47}$$

Next, define functions

$$\widehat{X}(t) = X(1) - \widetilde{X}(t) = \int_{t}^{1} \left((1-\tau)^{d_x} - 1_{\{\tau<0\}}(-\tau)^{d_x}\right) \mathrm{d}U(\tau) \tag{5.48}$$

and

$$\widehat{Y}(\tau) = Y(1) - \widetilde{Y}(\tau) = \int_{\tau}^{1} \left((1-t)^{d_y} - 1_{\{t<0\}}(-t)^{d_y}\right) \mathrm{d}W(t). \tag{5.49}$$

Setting $F_X = \widehat{X}$, for example, consider the following alternative representations of the iterated integral:

$$\int \widehat{X}\delta Y = \int_{-\infty}^{1}\int_{-\infty}^{1} 1_{\{\tau \geq t\}}\left((1-\tau)^{d_x} - 1_{\{\tau<0\}}(-\tau)^{d_x}\right)$$
$$\times\left((1-t)^{d_y} - 1_{\{t<0\}}(-t)^{d_y}\right)\mathrm{d}U(\tau)\mathrm{d}W(t)$$
$$= \int \widetilde{Y}\delta X + \mathrm{E}\left(\int \widehat{X}\delta Y\right). \tag{5.50}$$

The order of iteration is being swapped here, the indicator function in the second member setting the domain of integration to be the upper triangle of the plane $(-\infty, 1] \times (-\infty, 1]$. Since $\mathrm{E}\left(\int \widetilde{X}\delta Y\right) = 0$, the additional term in the third member of (5.50) equates the means of each side and can be viewed as representing the diagonal contribution to the double integral. Direct calculation using

$$\mathrm{E}\left(\mathrm{d}U(\tau)\mathrm{d}W(t)\right) = \begin{cases} \sigma_{uw}\mathrm{d}t, & t = \tau \\ 0, & \text{otherwise} \end{cases}$$

gives

$$\mathrm{E}\left(\int \widehat{X}\delta Y\right) = \sigma_{uw}\int_{-\infty}^{1}\left((1-t)^{d_x} - 1_{\{t<0\}}(-t)^{d_x}\right)\left((1-t)^{d_y} - 1_{\{t<0\}}(-t)^{d_y}\right)\mathrm{d}t$$
$$= \sigma_{uw}\left(\int_0^1 (1-t)^{d_x+d_y}\,\mathrm{d}t + \int_0^{\infty}\left((1+t)^{d_x} - t^{d_x}\right)\left((1+t)^{d_y} - t^{d_y}\right)\mathrm{d}t\right)$$
$$= \sigma_{uw}\Upsilon_{xy} \tag{5.51}$$

where the last equality is by (4.20). Given the first equality in (5.48) and the fact that $\mathrm{E}\left(\int \widetilde{X}\delta Y\right) = 0$, it can also be verified that $\mathrm{E}\left(\int \widehat{X}\delta Y\right) = \mathrm{E}(X(1)Y(1))$ so that (5.51) can be paired with the result already given in (4.21).

The final step in the argument is to add $\int \widetilde{X}\delta Y$ to each side of (5.50). Applying (5.47), (5.51), and the fact that $\int \widehat{X}\delta Y + \int \widetilde{X}\delta Y = X(1)Y(1)$ yields (5.39). By symmetry, to match (5.50) it is equally the case that

$$\int \widehat{Y}\delta X = \int \widetilde{X}\delta Y + \sigma_{uw}\Upsilon_{xy} \tag{5.52}$$

which gives the same result by addition of $\int \widetilde{Y}\delta X$ to both sides.

This completes the proof for the case $d_x \geq 0$ and $d_y \geq 0$. If $d_y < 0$ let $A_{xy}^*(t,s) + E_{yx}^*(t,s)$ replace $A_{xy}(t,s) + E_{yx}(t,s)$ in (5.40) and if $d_x < 0$, let $A_{yx}^*(s,t) + E_{xy}^*(s,t)$ replace $A_{yx}(s,t) + E_{xy}(s,t)$. That the argument does not depend in the signs of d_x and d_y follows from (5.43) and (5.44). ∎

In view of (5.39) and the fact $\Upsilon_{yx} = \Upsilon_{xy}$, when $d_x + d_y \leq 0$ it might appear harmless to assume the formal definition

$$\int_0^1 XdY = \Xi_{xy} + \tfrac{1}{2}\sigma_{uw}\Upsilon_{xy} \tag{5.53}$$

so that equation (5.38) is generally true. A single additional observation may provide food for thought on this point. In the case $d_x = d_y = 0$, so that $\Upsilon_{xy} = 1$ and $X = U$ and $Y = W$ are regular Brownian motions, $\Xi_{xy} = \Xi_{1xy}$ since $\Xi_{2xy} = 0$ and Ξ_{1xy} is the Itô integral of U with respect to W on the interval $[0, 1]$. It follows that (5.53) is the Stratonovich integral of U with respect to W. Letting $\{t_0, \dots, t_n\}$ be a partition of $[0, 1]$ with $\max\{t_{i+1} - t_i\} = O(1/n)$, the Stratonovich integral is the mean-square limit as $n \to \infty$ of the sum

$$\sum_{i=0}^{n-1}\left(\frac{U_{t_i} + U_{t_{i+1}}}{2}\right)(W_{t_{i+1}} - W_{t_i}) = \sum_{i=0}^{n-1} U_{t_i}(W_{t_{i+1}} - W_{t_i})$$
$$+ \tfrac{1}{2}\sum_{i=0}^{n-1}(U_{t_{i+1}} - U_{t_i})(W_{t_{i+1}} - W_{t_i}).$$

The limit of the first right-hand side term is the Itô integral and the second term converges in mean square to $\frac{1}{2}\sigma_{uw}$ by the law of large numbers.

Sources that may provide a deeper analysis of these issues include [44], [65], and also [25].

Chapter 6
Weak Convergence of Integrals

This chapter explores the conditions under which

$$(X_n, Y_n, G_n - \mathrm{E}(G_n)) \xrightarrow{\mathrm{d}} (X, Y, \Xi) \tag{6.1}$$

denoting joint weak convergence in $D^2_{[0,1]} \times \mathbb{R}$ where $D^2_{[0,1]}$ denotes the space of càdlàg pairs on the unit interval equipped with the Skorokhod topology, as explained on page 62. In (6.1), X_n and Y_n are the normalized partial sums of fractional processes x_i and y_i defined for the exemplar case by (2.28) with respective shock variances σ_u^2 and σ_w^2 and covariance σ_{uw}, while X and Y are fractional Brownian motions as in (2.1). It is understood that G_n and Ξ are the objects that were written in §5.5 as G_n^{xy} and Ξ_{xy}, with the additional decorations now omitted to minimize clutter.

The joint convergence of the triple specified in (6.1) is a requisite for arguments depending on application of the continuous mapping theorem to functionals of the components. The scalar $G_n - \mathrm{E}(G_n) = G_{1n} + G_{3n}$ can be embedded in $D_{[0,1]}$ by defining a process on $[0, 1]$ to equal it at every point. Then by Cramér-Wold (Theorem **3.19**), it is required that arbitrary linear combinations of $(X_n, Y_n, G_n - \mathrm{E}(G_n))$ converge to the corresponding combinations of the limit processes. The marginal weak limits for the first two members of (6.1) follow in effect from Theorem **3.21** under the conditions specified. If $G_{1n} \to_{\mathrm{d}} \Xi_1$ and $G_{3n} \to_{\mathrm{d}} \Xi_3$ where the limit random variables Ξ_1 and Ξ_3 are the stochastic integrals in (5.23) and (5.29), and $\Xi = \Xi_1 + \Xi_3$, the continuous mapping theorem would then yield the weak limit for the third element of (6.1).

6.1 More Fractional Asymptotics

In Chapters 4 and 5, formulae have been derived that intuitively correspond with the limiting form of the fractional covariance, respectively the mean process and the mean deviations, as the sample size n increases. It now has to be shown that these components do converge weakly to the identified limit distributions.

Asymptotics for Fractional Processes. James Davidson, Oxford University Press. © James Davidson (2025).
DOI: 10.1093/9780198955207.003.0006

The following mechanical lemmas provide bounds for the limit formulae.

6.1 Lemma $K(n)^{-1}\left|a_{n[nt][ns]}\right| \leq \bar{A}(t,s) + o(1)$ as $n \to \infty$, where

$$\bar{A}(t,s) = \left|(1-t)^{d_y} - 1_{\{t<0\}}(-t)^{d_y}\right| \left|(g-s)^{d_x} - 1_{\{s<0\}}(-s)^{d_x}\right| \tag{6.2}$$

with $g = t$ if $d_x < 0$ and $s \geq 0$, and $g = 1$ otherwise.

Proof Putting $[nt]$ for m and $[ns]$ for p in formula (5.5) and defining g by $[ng] = l + [nt]$,

$$\begin{aligned}\frac{\left|a_{n[nt][ns]}\right|}{K(n)} &= \frac{1}{K(n)}\left|\sum_{l=\max\{0,1-[nt]\}}^{n-1-[nt]} c_l \left(\sum_{j=\max\{0,1-[ns]\}}^{l+[nt]-[ns]} b_j\right)\right| \\ &\leq \frac{1}{K(n)}\left|\sum_{l=\max\{0,1-[nt]\}}^{n-1-[nt]} c_l\right| \max_{\max\{0,1-[nt]\}\leq l\leq n-1-[nt]}\left|\sum_{j=\max\{0,1-[ns]\}}^{l+[nt]-[ns]} b_j\right| \\ &= \left|(1-t)^{d_y} - 1_{\{t<0\}}(-t)^{d_y}\right| \\ &\qquad \times \max_{\max\{0,t\}\leq g\leq 1}\left|(g-s)^{d_x} - 1_{\{s<0\}}(-s)^{d_x}\right| + o(1). \end{aligned} \tag{6.3}$$

When $d_x \geq 0$, $(g-s)^{d_x}$ is monotone nondecreasing in g, and is maximized over $[0,1]$ at $g = 1$. When $d_x < 0$, $(g-s)^{d_x}$ is monotone decreasing in g, and if $s \geq 0$ so that $t \geq 0$, the maximum in (6.3) is achieved at $g = t$. On the other hand, if $d_x < 0$ and $s < 0$ then $\left|(g-s)^{d_x} - (-s)^{d_x}\right|$ is maximized at $g = 1$. ∎

Since $\bar{A}(t,s)$ bounds both $A(t,s)$ in (5.10) and $A^*(t,s)$ in (5.32), results that depend on bounding $\bar{A}(t,s)$ hold both for $|d_x| < \frac{1}{2}$ and for $|d_y| < \frac{1}{2}$.

6.2 Lemma $K(n)^{-1}\left|e_{n[ns][nt]}\right| \leq \bar{E}(s,t) + o(1)$ where

$$\bar{E}(s,t) = \left|(1-s)^{d_x} - 1_{\{s<0\}}(-s)^{d_x}\right| \left|(1-t)^{d_y} - 1_{\{t<0\}}(-t)^{d_y}\right|. \tag{6.4}$$

Proof From (5.8), for $s \geq t$, defining g by $[ng] = j + [ns]$,

$$\begin{aligned}\frac{\left|e_{n[ns][nt]}\right|}{K(n)} &= \frac{1}{K(n)}\left|\sum_{j=\max\{0,1-[ns]\}}^{n-1-[ns]} b_j \left(\sum_{l=j+[ns]+1-[nt]}^{n-[nt]} c_l\right)\right| \\ &\leq \frac{1}{K(n)}\left|\sum_{j=\max\{0,1-[ns]\}}^{n-1-[ns]} b_j\right| \max_{\max\{0,1-[ns]\}\leq j\leq n-1-[ns]}\left|\sum_{l=j+[ns]+1-[nt]}^{n-[nt]} c_l\right| \\ &= \left|(1-s)^{d_x} - 1_{\{s<0\}}(-s)^{d_x}\right| \\ &\qquad \times \max_{\max\{0,s\}\leq g\leq 1}\left|(1-t)^{d_y} - (g-t)^{d_y}\right| + o(1). \end{aligned}$$

First, suppose $s \geq 0$. Whether $d_y \geq 0$ or $d_y < 0$, $|(1-t)^{d_y} - (g-t)^{d_y}|$ is maximized over $[s, 1]$ at $g = s$. In the case $s < 0$ the same considerations apply, but the extremum over $[0, 1]$ is at $g = 0$ in each case. The proof is completed by noting that for any $s \in [t, 1]$ and d_y of either sign,

$$\left|(1-t)^{d_y} - (\max\{0, s\} - t)^{d_y}\right| \leq \left|(1-t)^{d_y} - 1_{\{t<0\}}(-t)^{d_y}\right|. \quad \blacksquare$$

In the same way as for $\bar{A}(t, s)$, results that work by bounding $\bar{E}(s, t)$ hold for $|d_x| < \frac{1}{2}$ and $|d_y| < \frac{1}{2}$.

To address the fact that the partial sums G_{1n} and G_{3n} have an infinite number of terms, the approach as in Chapter 3 is to split each sum into a block of order $n(N+1)$ and a remainder, where the remainder is of small order in L_2-norm as $N \to \infty$; not as $n \to \infty$, note, because N can and does characterize the limit distribution.

Because of the form of the sums, there are in practice two remainder terms to consider. For the chosen $N \in \mathbb{N}$, first define

$$G_{1n}^N = \frac{1}{\sqrt{n}} \sum_{m=-nN}^{n-1} q_{nm}^N w_{m+1} \tag{6.5}$$

where

$$q_{nm}^N = \frac{1}{\sqrt{n}K(n)} \sum_{p=-nN}^{m} a_{nmp} u_p. \tag{6.6}$$

Then (5.3) may be rearranged as

$$G_{1n} = G_{1n}^N + \frac{1}{\sqrt{n}} \sum_{m=-nN}^{n-1} (q_{nm} - q_{nm}^N) w_{m+1} + \frac{1}{\sqrt{n}} \sum_{m=-\infty}^{-nN-1} q_{nm} w_{m+1}. \tag{6.7}$$

Similarly, define

$$G_{3n}^N = \frac{1}{\sqrt{n}} \sum_{p=-nN}^{n-1} h_{np}^N u_p \tag{6.8}$$

and

$$h_{np}^N = \frac{1}{\sqrt{n}K(n)} \sum_{m=-nN}^{p-1} e_{npm} w_m \tag{6.9}$$

and so for G_{3n} in (5.6) write

$$G_{3n} = G_{3n}^N + \frac{1}{\sqrt{n}} \sum_{p=-nN}^{n-1} (h_{np} - h_{np}^N) u_p + \frac{1}{\sqrt{n}} \sum_{p=-\infty}^{-nN-1} h_{np} u_p. \tag{6.10}$$

It has to be shown that the limits of the last two terms on the right-hand sides of (6.7) and (6.10) as $n \to \infty$ can each be made as small as desired in L_2-norm by taking N large enough.

The next pair of theorems bound the remainders $G_{1n} - G_{1n}^N$ and $G_{3n} - G_{3n}^N$, defined respectively in (6.7) and (6.10).

6.3 Theorem Under Assumption **4.1**,

(i) $\lim_{n\to\infty} n^{-1}\mathrm{E}\left(\sum_{m=-nN}^{n-1}(q_{nm} - q_{nm}^N)w_{m+1}\right)^2 = O(N^{2d_x-1})$

(ii) $\lim_{n\to\infty} n^{-1}\mathrm{E}\left(\sum_{m=-\infty}^{-nN-1} q_{nm}w_{m+1}\right)^2 = O(N^{2(d_x+d_y-1)})$.

Proof For part (i),

$$\begin{aligned}
\frac{1}{n}\mathrm{E}\left(\sum_{m=-nN}^{n-1}(q_{nm} - q_{nm}^N)w_{m+1}\right)^2 &= \frac{1}{n^2K(n)^2}\mathrm{E}\left(\sum_{m=-nN}^{n-1}\sum_{p=-\infty}^{-nN-1} a_{nmp}u_pw_{m+1}\right)^2 \\
&= \sigma_u^2\sigma_w^2\frac{1}{n^2K(n)^2}\sum_{m=-nN}^{n-1}\sum_{p=-\infty}^{-nN-1} a_{nmp}^2 \\
&\le \sigma_u^2\sigma_w^2\int_{-N}^{1}\int_{-\infty}^{-N}\bar{A}^2(t,s)\mathrm{d}s\mathrm{d}t + o(1) \qquad (6.11)
\end{aligned}$$

as $n \to \infty$. Applying Lemma **6.1**, with $g = 1$ since $s < 0$,

$$\begin{aligned}
\int_{-N}^{1}\int_{-\infty}^{-N}\bar{A}^2(t,s)\mathrm{d}s\mathrm{d}t &= \int_0^{N+1}\left((1+t)^{d_y} - t^{d_y}\right)^2\mathrm{d}t\int_N^\infty\left((1+s)^{d_x} - s^{d_x}\right)^2\mathrm{d}s \\
&= O(N^{2d_x-1}). \qquad (6.12)
\end{aligned}$$

For part (ii), similarly,

$$\begin{aligned}
\frac{1}{n}\mathrm{E}\left(\sum_{m=-\infty}^{-nN-1} q_{nm}w_{m+1}\right)^2 &= \sigma_u^2\sigma_w^2\frac{1}{n^2K(n)^2}\sum_{m=-\infty}^{-nN-1}\sum_{p=-\infty}^{m} a_{nmp}^2 \\
&\le \sigma_u^2\sigma_w^2\int_{-\infty}^{-N}\int_{-\infty}^{t}\bar{A}^2(t,s)\mathrm{d}s\mathrm{d}t + o(1) \qquad (6.13)
\end{aligned}$$

where

$$\begin{aligned}
\int_{-\infty}^{-N}\int_{-\infty}^{t}\bar{A}^2(t,s)\mathrm{d}s\mathrm{d}t &= \int_N^\infty\left((1+t)^{d_y} - t^{d_y}\right)^2\int_t^\infty\left((1+s)^{d_x} - s^{d_x}\right)^2\mathrm{d}s\mathrm{d}t \\
&\ll \int_N^\infty\left((1+t)^{d_y} - t^{d_y}\right)^2 t^{2d_x-1}\mathrm{d}t \\
&= O(N^{2(d_y+d_x-1)}). \quad \blacksquare \qquad (6.14)
\end{aligned}$$

6.4 Theorem Under Assumption **4.1**,

(i) $\lim_{n\to\infty} n^{-1}\mathrm{E}\left(\sum_{p=-nN}^{n-1}(h_{np} - h_{np}^N)u_p\right)^2 = O(N^{2d_y-1})$

(ii) $\lim_{n\to\infty} n^{-1}\mathrm{E}\left(\sum_{p=-\infty}^{-nN-1} h_{np}u_p\right)^2 = O(N^{2(d_x+d_y-1)})$.

Proof For part (i), following the pattern of (6.11) and applying Lemma **6.2**,

$$\frac{1}{n}\mathrm{E}\left(\sum_{p=-nN}^{n-1}(h_{np} - h_{np}^N)u_p\right)^2 \le \sigma_w^2\sigma_u^2\int_{-N}^{1}\int_{-\infty}^{-N}\bar{E}^2(s,t)\mathrm{d}t\mathrm{d}s + o(1)$$
$$= O(N^{2d_y-1}). \tag{6.15}$$

For part (ii), as in (6.14),

$$\frac{1}{n}\mathrm{E}\left(\sum_{p=-\infty}^{-nN-1} h_{np}u_p\right)^2 \le \sigma_w^2\sigma_u^2\int_{-\infty}^{-N}\int_{-\infty}^{s}\bar{E}^2(s,t)\mathrm{d}t\mathrm{d}s + o(1)$$
$$= O(N^{2(d_x+d_y-1)}). \quad \blacksquare \tag{6.16}$$

For the subsequent analysis the normalized sums $a_{nmp}/K(n)$ and $e_{npm}/K(n)$ must be shown to satisfy asymptotic continuity conditions. The following two theorems follow essentially the same lines so that the proofs can be read in conjunction, but there are some minor differences between the formulae. Since (6.4) is not dependent on a parameter like (6.2), some complications are avoided in that case.

6.5 Theorem Defining a_{nmp} as in (5.5),

(i) $\sup_{t\in(-\infty,1-\delta]}\limsup_n n^{-1}K(n)^{-2}\sum_{p=[nt]+1}^{[n(t+\delta)]} a_{n[n(t+\delta)]p}^2 = O(\delta^{\min\{1,2d_x+1\}})$.

(ii) $\sup_{t\in(-\infty,1-\delta]}\limsup_n n^{-1}K(n)^{-2}\sum_{p=-\infty}^{[nt]}(a_{n[n(t+\delta)]p} - a_{n[nt]p})^2 = O(\delta^{2d_x+1})$.

Proof To prove (i), write

$$\limsup_n \frac{1}{nK(n)^2}\sum_{p=[nt]+1}^{[n(t+\delta)]} a_{n[n(t+\delta)]p}^2 \le \int_t^{t+\delta}\bar{A}^2(t+\delta,s)\mathrm{d}s$$
$$= \left((1-t-\delta)^{d_y} - 1_{\{t+\delta<0\}}(-t-\delta)^{d_y}\right)^2$$
$$\times\int_t^{t+\delta}\left((g-s)^{d_x} - 1_{\{s<0\}}(-s)^{d_x}\right)^2\mathrm{d}s.$$

If $t \geq 0$ and $d_x < 0$, Lemma **6.1** sets $g = t + \delta$ and

$$\int_t^{t+\delta} (t + \delta - s)^{2d_x} ds = \frac{\delta^{2d_x+1}}{2d_x + 1}.$$

Otherwise, $g = 1$. If $t \geq 0$ then according to the mean value theorem,

$$\begin{aligned}\int_t^{t+\delta} (1 - s)^{2d_x} ds &= \frac{(1 - t)^{2d_x+1} - (1 - t - \delta)^{2d_x+1}}{2d_x + 1} \\ &= O(\delta).\end{aligned} \tag{6.17}$$

If $t < 0$ then with δ sufficiently small, $t + \delta \leq 0$. Similarly to (6.17),

$$\begin{aligned}\int_t^{t+\delta} \left((1 - s)^{d_x} - (-s)^{d_x}\right)^2 ds &\leq 2\left(\int_t^{t+\delta} (1 - s)^{2d_x} ds + \int_t^{t+\delta} (-s)^{2d_x} ds\right) \\ &= O(\delta).\end{aligned}$$

The indicated bound holds uniformly with respect to $t \in (-\infty, 1)$.

To prove (ii) consider the formula in (5.5) for the case where m is replaced by $m + q$ where $0 < q < n - m$. The formula can be decomposed as

$$a_{n,m+q,p} = \left(\sum_{l=\max\{1-m-q,0\}}^{\max\{1-m,0\}-1} + \sum_{l=\max\{1-m,0\}}^{n-m-q-1}\right) c_l \left(\sum_{j=\max\{1-p,0\}}^{l+m-p} + \sum_{j=l+m-p+1}^{l+m+q-p}\right) b_j$$

whereas

$$a_{nmp} = \left(\sum_{l=\max\{1-m,0\}}^{n-m-q-1} + \sum_{l=n-m-q}^{n-m-1}\right) c_l \sum_{j=\max\{1-p,0\}}^{l+m-p} b_j.$$

After cancelling equal and opposite-signed terms,

$$a_{n,m+q,p} - a_{nmp} = D_{1n}(q, m, p) + D_{2n}(q, m, p) - D_{3n}(q, m, p) \tag{6.18}$$

where

$$\begin{aligned}D_{1n}(q, m, p) &= \sum_{l=\max\{1-m-q,0\}}^{\max\{1-m,0\}-1} c_l \sum_{j=\max\{1-p,0\}}^{l+m-p} b_j \\ D_{2n}(q, m, p) &= \sum_{l=\max\{1-m-q,0\}}^{n-m-q-1} c_l \sum_{j=l+m-p+1}^{l+m+q-p} b_j \\ D_{3n}(q, m, p) &= \sum_{l=n-m-q}^{n-m-1} c_l \sum_{j=\max\{1-p,0\}}^{l+m-p} b_j.\end{aligned}$$

D_{1n} vanishes unless $m \le 0$. Using arguments analogous to (6.3), writing $q = [n\delta]$ for $0 < \delta < 1 - t$ the inequalities

$$\frac{|D_{jn}([n\delta], [nt], [ns])|}{K(n)} \le \bar{D}_j(\delta, t, s) + o(1)$$

hold for $j = 1, 2$, and 3, where

$$\bar{D}_1(\delta, t, s) = 1_{\{t<0\}}|(-t)^{d_y} - 1_{\{\delta+t<0\}}(-\delta - t)^{d_y}| \\ \times \max_{\max\{-t-\delta,0\}\le g_1\le \max\{-t,0\}} |(g_1 + t - s)^{d_x} - 1_{\{s<0\}}(-s)^{d_x}| \tag{6.19}$$

$$\bar{D}_2(\delta, t, s) = |(1 - t - \delta)^{d_y} - 1_{\{t+\delta<0\}}(-t - \delta)^{d_y}| \\ \times \max_{\max\{-t-\delta,0\}\le g_2\le 1-t-\delta} |(g_2 + \delta - s)^{d_x} - (g_2 - s)^{d_x}| \tag{6.20}$$

$$\bar{D}_3(\delta, t, s) = |(1 - t)^{d_y} - (1 - t - \delta)^{d_y}| \\ \times \max_{1-t-\delta\le g_3\le 1-t} |(g_3 + t - s)^{d_x} - (-s)^{d_x}|. \tag{6.21}$$

In view of (6.18),

$$\limsup_{n\to\infty} \frac{1}{nK(n)^2} \sum_{p=-\infty}^{[nt]} (a_{n[n(t+\delta)]p} - a_{n[nt]p})^2 \le 3\sum_{j=1}^{3} \int_{-\infty}^{t} \bar{D}_j^2(\delta, t, s)\mathrm{d}s.$$

The squares of the second factors of $\bar{D}_1$ in (6.19) and $\bar{D}_3$ in (6.21) are integrable with respect to s. Considering the squared first factors, applying the mean value theorem shows that $\int_{-\infty}^{t} \bar{D}_j^2(\delta, t, s)\mathrm{d}s = O(\delta^2)$ for $j = 1$ and $j = 3$. In the case of $\bar{D}_1$, $1_{\{t+\delta<0\}} = 0$ at the same time as $1_{\{t<0\}} = 1$ only if $-\delta < t < 0$, so this does not happen whenever δ is small enough. In the case of (6.20), the integral of the squared second factor of $\bar{D}_2$ needs to be bounded. To do this put g_2^* for the maximizing value of g_2 and then make the change of variable $x = (g_2^* - s)/\delta$ in the integral, and so verify that $\int_{-\infty}^{t} \bar{D}_2^2(\delta, t, s)\mathrm{d}s = O(\delta^{2d_x+1})$. These orders of magnitude are all independent of t, and (ii) follows. ∎

6.6 Theorem Defining e_{npm} as in (5.8),

(i) $\sup_{s\in(-\infty,1-\delta]} \limsup_{n} n^{-1}K(n)^{-2} \sum_{m=[ns]+1}^{[n(s+\delta)]} e_{n[ns]m}^2 = O(\delta)$.

(ii) $\sup_{s\in(-\infty,1-\delta]} \limsup_{n} n^{-1}K(n)^{-2} \sum_{m=-\infty}^{[ns]} (e_{n[n(s+\delta)]m} - e_{n[ns]m})^2 = O(\delta^{2d_y+1})$.

Proof The proof of (i) is simpler than in Theorem **6.5**(i) in view of (6.4). The result

$$\limsup_n \frac{1}{nK(n)^2} \sum_{m=[ns]+1}^{[n(s+\delta)]} e^2_{n[n(s+\delta)]m} \le \int_s^{s+\delta} \bar{E}^2(s+\delta, t)\mathrm{d}t = O(\delta)$$

holds similarly to (6.17).

For part (ii), the decomposition analogous to (6.18) for $0 < q < n - p$ is

$$e_{n,p+q,m} = \left(\sum_{j=\max\{0,1-p-q\}}^{\max\{0,1-p\}-1} + \sum_{j=\max\{0,1-p\}}^{n-1-p-q} \right) b_j \sum_{l=j+p+q+1-m}^{n-m} c_l$$

$$e_{npm} = \left(\sum_{j=\max\{0,1-p\}}^{n-1-p-q} + \sum_{n-p-q}^{n-1-p} \right) b_j \left(\sum_{l=j+p-m}^{j+p+q-m} + \sum_{l=j+p+q+1-m}^{n-m} \right) c_l.$$

It can be verified that in this case

$$e_{n,p+q,m} - e_{npm} = D_{1n}(q,p,m) - D_{2n}(q,p,m) - D_{3n}(q,p,m)$$

where

$$D_{1n}(q,p,m) = \sum_{j=\max\{0,1-p-q\}}^{\max\{0,1-p\}-1} b_j \sum_{l=j+p+q+1-m}^{n-m} c_l$$

$$D_{2n}(q,p,m) = \sum_{n-p-q}^{n-1-p} b_j \sum_{l=j+p+q+1-m}^{n-m} c_l$$

$$D_{3n}(q,p,m) = \sum_{j=\max\{0,1-p\}}^{n-1-p} b_j \sum_{l=j+p-m}^{j+p+q-m} c_l.$$

D_{1n} vanishes unless $p \le 0$. For $0 < \delta < 1 - s$ the inequalities

$$\frac{|D_{jn}([n\delta],[ns],[nt])|}{K(n)} \le \bar{D}_j(\delta, s, t) + o(1)$$

hold for $j = 1, 2$, and 3, where

$$\begin{aligned}\bar{D}_1(\delta,s,t) &= 1_{\{s<0\}}\left|(-s)^{d_x} - 1_{\{s+\delta<0\}}(-s-\delta)^{d_x}\right| \\ &\quad \times \max_{\max\{-s-\delta,0\}\le g_1 \le \max\{-s,0\}} \left|(1-t)^{d_y} - (g_1+s+\delta-t)^{d_y}\right| \end{aligned} \tag{6.22}$$

$$\bar{D}_2(\delta, s, t) = |(1-s)^{d_x} - (1-s-\delta)^{d_x}| \times \max_{1-s-\delta \le g_2 \le 1-s} |(1-t)^{d_y} - (g_2 + s + \delta - t)^{d_y}| \quad (6.23)$$

$$\bar{D}_3(\delta, s, t) = |(1-s)^{d_x} - 1_{\{s<0\}}(-s)^{d_x}| \times \max_{\max\{0,-s\} \le g_3 \le 1-s} |(g_3 + t + \delta - s)^{d_y} - (g_3 + t - s)^{d_y}| \quad (6.24)$$

and

$$\limsup_{n\to\infty} \frac{1}{nK(n)^2} \sum_{m=-\infty}^{[ns]} (e_{n[n(s+\delta)]m} - e_{n[ns]m})^2 \le 3 \sum_{j=1}^{3} \int_{-\infty}^{s} \bar{D}_j^2(\delta, s, t)\mathrm{d}t.$$

The analysis is now very similar to the proof of Theorem **6.5**(ii). The squared second factors of $\bar{D}_1$ in (6.22) and $\bar{D}_2$ in (6.23), are integrable with respect to t. The mean value theorem applied to the first factors therefore gives $\int_{-\infty}^{t} \bar{D}_j^2(\delta, s, t)\mathrm{d}s = O(\delta^2)$ for $j = 1$ and $j = 2$ where in the case of $\bar{D}_1$, $1_{\{s+\delta<0\}} = 0$ and $1_{\{s<0\}} = 1$ only if $\delta > -s$. In the case of (6.24), put g_3^* for the maximizing value of g_3, make the change of variable $x = (g_3^* + t - s)/\delta$ and so verify that $\int_{-\infty}^{s} \bar{D}_3^2(\delta, s, t)\mathrm{d}t = O(\delta^{2d_y+1})$. ∎

6.2 The Main Result

According to (6.7) and (6.10), the finite sums G_{1n}^N and G_{3n}^N differ from G_{1n} and G_{3n} by tail components that are shown, respectively by Theorems **6.3** and **6.4**, to be negligible in L_2-norm when N is large enough. The sequences $\{w_m\}$ and $\{u_p\}$ as specified in Assumption **4.1** are mapped to càdlàg processes on the real line segment by writing $w_n^N(t) = w_{[nt]}$, $t \in [-N, 1]$ and $u_n^N(s) = u_{[ns]}$, $s \in [-N, 1]$. In the same way, the discrete arrays $\{q_{nm}^N\}$ from (6.6) and $\{h_{np}^N\}$ from (6.9), are mapped to $D_{[-N,1]}$ by the assignments

$$q_n^N(t) = q_{n[nt]}^N, \; t \in [-N, 1] \quad (6.25)$$

and

$$h_n^N(s) = h_{n[ns]}^N, \; s \in [-N, 1]. \quad (6.26)$$

The counterpart truncations for the putative limit processes (5.23) and (5.29) are $\Xi_1^N = \int_{-N}^{1} Q^N(t)\mathrm{d}W(t)$, where $Q^N(t) = \int_{-N}^{t} A(t,s)\mathrm{d}U(s)$, and also $\Xi_3^N = \int_{-N}^{1} H^N(s)\mathrm{d}U(s)$ where $H^N(s) = \int_{-N}^{s} E(s,t)\mathrm{d}W(t)$. By analogy with (2.19), Theorems **6.3** and **6.4** provide the bounds required to show the orders of magnitude of the respective remainders, which are as follows.

6.7 Corollary Under Assumption **4.1**,

$$E(\Xi_1 - \Xi_1^N)^2 = O(N^{\max\{2d_x-1, 2(d_x+d_y-1)\}}).$$

Proof Lemma **6.1** gives $A^2(t,s) \leq \bar{A}^2(t,s)$ and $A^{*2}(t,s) \leq \bar{A}^2(t,s)$ and hence

$$\begin{aligned} E(\Xi_1 - \Xi_1^N)^2 &= E\left(\int_{-\infty}^{-N} Q(t)dW(t) + \int_{-N}^{1} (Q(t) - Q^N(t))dW(t)\right)^2 \\ &\leq 2\sigma_u^2\sigma_w^2\left(\int_{-\infty}^{-N}\int_{-\infty}^{t} \bar{A}^2(t,s)dsdt + \int_{-N}^{1}\int_{-\infty}^{-N} \bar{A}^2(t,s)dsdt\right). \end{aligned}$$

The result follows by (6.12) and (6.14). ∎

6.8 Corollary Under Assumption **4.1**,

$$E(\Xi_3 - \Xi_3^N)^2 = O(N^{\max\{2d_y-1, 2(d_x+d_y-1)\}}).$$

Proof Similarly, by Lemma **6.2**,

$$\begin{aligned} E(\Xi_3 - \Xi_3^N)^2 &= E\left(\int_{-\infty}^{-N} H(s)dU(s) + \int_{-N}^{1} (H(s) - H^N(s))dU(s)\right)^2 \\ &\leq 2\sigma_u^2\sigma_w^2\left(\int_{-\infty}^{-N}\int_{-\infty}^{s} \bar{E}^2(s,t)dtds + \int_{-N}^{1}\int_{-\infty}^{-N} \bar{E}^2(s,t)dtds\right). \end{aligned}$$

The result follows by (6.15) and (6.16). ∎

The weak convergence results of this chapter are given for the finite-domain processes that are indicated with superscript N. Corollaries **6.7** and **6.8** provide the means to link these results with limit processes Ξ_1 and Ξ_3. Statements of the form '$G_{1n} \to_d \Xi_1$' and '$G_{3n} \to_d \Xi_3$' are to be understood as a compact expression of these successive convergences, in distribution and then in mean square.

In addition to Assumption **4.1**, a further assumption is needed for the weak convergences of q_n^N and h_n^N.

6.9 Assumption $d_x + d_y > -\frac{1}{2}$ and

(a) $\{u_i\}_{i=-\infty}^{\infty}$ is L_r-bounded for $r \geq \max\{2, 1/(\frac{1}{2}+d_x), 1/(\frac{1}{2}+d_x+d_y)\}$

(b) $\{w_i\}_{i=-\infty}^{\infty}$ is L_r-bounded for $r \geq \max\{2, 1/(\frac{1}{2}+d_y), 1/(\frac{1}{2}+d_x+d_y)\}$. □

This assumption incorporates Assumption **3.1** for u_i and the counterpart condition for w_i, so that Theorem **3.2** applies for both variables, but there are also new conditions to handle the additional weak convergences arising in this chapter. These can be seen to reflect the same concerns that motivated Assumption **3.1** and are needed for the results appearing in §6.4.

The convergence of the third element of (6.1) is a consequence of the following joint weak convergence in the space $D^4_{[-N,1]} \times \mathbb{R}^2$, where $D^4_{[-N,1]}$ is endowed with the Skorokhod topology. Pages 23 and 62 provide some related details here, noting that the same considerations arise when the domain of the functions is generalized beyond $[0, 1]$.

6.10 Theorem Under Assumptions **4.1** and **6.9**,

$$\left(w_n^N, q_n^N, u_n^N, h_n^N, G_{1n}^N, G_{3n}^N \right) \xrightarrow{d} \left(W^N, Q^N, U^N, H^N, \Xi_1^N, \Xi_3^N \right) \tag{6.27}$$

where (W^N, Q^N, U^N, H^N) are Gaussian, a.s. continuous, and adapted to a common filtration $\{\mathcal{F}(t), t \in [-N, 1]\}$ with respect to which W^N and U^N are martingales. □

The convergence to a.s. continuous limits of the first four elements in (6.27) is a property called upon in the proof of the last two convergences, hence the joint treatment of these elements is unavoidable. The cases of w_n^N and u_n^N are standard results, the conditions of Theorem **3.2** being satisfied directly with $d = 0$. W^N and U^N are Brownian motions on $[-N, 1]$ with $W^N(0) = U^N(0) = 0$ and variances $\mathrm{E}(W^N(1)^2) = \sigma_w^2$ and $\mathrm{E}(U^N(1)^2) = \sigma_u^2$. These processes contribute to the covariances only in increment form, so their location is arbitrary and can be assumed as such without loss of generality.

Requiring special attention by contrast are the convergences in distribution of q_n^N in (6.25) and h_n^N in (6.26). These partial sum processes evolve by re-weighting existing increments as well as by appending new ones. However, for each fixed value of t, $q_n^N(t)$ and $h_n^N(t)$ are the terminal points of partial sums defined on $[-N, t]$, having heteroscedastic but independent increments. These are shown to have Gaussian limits by a conventional central limit theorem on the lines of Lemma **3.4**.

For example, consider the process

$$\widetilde{Q}^N(t, \tau) = \int_{-N}^{\tau} A(t, s)\mathrm{d}U^N(s), \ \ \tau \in (-N, t).$$

This is a variance-transformed Brownian motion depending on a constant t and having increment weights $A(t, s)$ that depend on t but otherwise only on the time s. For each t, this is shown by conventional means to be a Gaussian process with known variance function. The process Q^N itself is then defined for each t by setting $\tau = t$, so that

$$Q^N(t) = \widetilde{Q}^N(t,t), \ t \in (-N, 1). \tag{6.28}$$

The process H^N is defined analogously. These are the limits as $n \to \infty$ of empirical processes q_n^N and h_n^N that are likewise constructed from the terminal points of partial sum processes $q_n^N(t)$ and $h_n^N(t)$ for $t \in (-N, 1]$. The proof of the functional central limit theorem is completed by showing stochastic equicontinuity for q_n^N and h_n^N, so as to establish that the limits Q^N and H^N so defined are a.s. continuous. This can be done by some relatively minor variations of previously established arguments.

The proofs for the cases of q_n^N and h_n^N are closely parallel and only that for q_n^N is given in full. This is as follows.

6.11 Theorem Under Assumptions **4.1** and **6.9**(a), $q_n^N \to_d Q^N$ where the limit is an a.s. continuous Gaussian process on the interval $(-N, 1)$. □

As with Theorem **3.2**, the proof of this theorem falls into two parts occupying the next two sections, Lemma **6.12** in §6.3 showing the limiting Gaussianity while Lemma **6.14** in §6.4 proves the stochastic equicontinuity. In each case, the arguments closely resemble those given for the FCLT in Chapter 3.

6.3 Finite Dimensional Distributions

If $q_n^N(t)$ is the normalized sum in (6.25), for given t, the following lemma shows that under Assumption **4.1** its limit in distribution as $n \to \infty$ is Gaussian with variance

$$V^N(t) = \mathrm{E}(Q^N(t)^2) = \begin{cases} \sigma_u^2 \int_{-N}^t A^2(t,s)ds, & d_y > 0 \\ \sigma_u^2 \int_{-N}^t A^{*2}(t,s)ds, & d_y < 0. \end{cases} \tag{6.29}$$

The argument is similar to that of Lemma **3.4** since the terms of the sum are again independent, but the form of the moving average weights is different.

6.12 Lemma Under Assumption **4.1**, $q_n^N(t) \to_d \mathrm{N}(0, V^N(t))$ for each $t \in (-N, 1)$.

Proof Similarly to the decomposition of (3.2), it is convenient to split the sums into two components. Recalling the representations in (6.25) and (6.6), set

$$q_n^N(t) = q_{1n}(t) + q_{2n}^N(t) \tag{6.30}$$

where $q_{1n}(t)$ has the terms labelled $p = 1, \ldots, [nt]$ in the cases with $t > 0$ and $q_{1n}(t) = 0$ when $t \leq 0$, while $q_{2n}^N(t)$ has the terms labelled $p = 1 - nN, \ldots, \min\{0, [nt]\}$. The cases $d_y > 0$ and $d_y < 0$ are treated separately, the logic being the same but the formulae different.

Case 1: $d_y > 0$.

Decompose the function $A(t, s)$ in (5.10) as

$$A(t,s) = 1_{\{s>0\}}A_1(t,s) + 1_{\{s\leq 0\}}A_2(t,s) \tag{6.31}$$

where A_1 is the first term of (5.10). According to Lemma **5.2**, since $t \geq s$ the moving average weights in $q_{1n}(t)$ have the form $K(n)^{-1}a_{n[nt][ns]} = A_1(t,s) + o(1)$, where, using the representation in (5.19) for $s < t$,

$$A_1(t,s) = (1-t)^{d_y}(t-s)^{d_x}Z_1^A(t,s). \tag{6.32}$$

Set $p = [ns]$ and so write the array of moving average coefficients from (6.6) as $\{c_{np}\}_{p=1}^{[nt]}$ where according to (6.32) for $s < t$,

$$c_{np} = \frac{|a_{n,[nt],[nt]-p+1}|}{\sqrt{n}K(n)} \sim (1-t)^{d_y}Z_1^A(t,s)\frac{([nt]-p)^{d_x}}{n^{d_x+1/2}}. \tag{6.33}$$

The slowly varying components cancel in the limit. The point $s = t$ is excluded from the equivalence in (6.33) since Z_1^A diverges there if $d_x > 0$ and equals zero if $d_x < 0$.

Apply the Lindeberg condition test of Lemma **3.4** with c_{np}^2 defined in (6.33) replacing $a_{ni}^2/\kappa(n)^2$ in (3.8). Similarly to (3.9),

$$\sum_{p=1}^{[nt]} c_{np}^2 \mathrm{E}(u_p^2 1_{\{|c_{np}u_p|>\varepsilon\}}) \leq \max_{1\leq p\leq [nt]} \mathrm{E}(u_p^2 1_{\{|u_p|>\varepsilon/c_{np}\}})\sum_{p=1}^{[nt]} c_{np}^2 \tag{6.34}$$

where $\sum_{p=1}^{[nt]} c_{np}^2 = O(1)$ according to (6.33). In the case $d_x > 0$ the maximum in question is at $p = 1$ and $c_{n1} = O(n^{-1/2})$. If $d_x < 0$ on the other hand, the maximum in (6.34) is at $p = [nt]$. The representation (6.33) is not useful here, but instead refer directly to (5.5) which shows that

$$\begin{aligned} a_{n[nt][nt]} &= c_0 b_0 + c_1(b_0 + b_1) + \cdots + c_{n-1-[nt]}(b_0 + \cdots + b_{n-1-[nt]}) \\ &= O(n^{d_y} L_x(n) L_y(n)) \end{aligned}$$

so $c_{n[nt]} = O(n^{-d_x-1/2})$. Theorem **A.5** of Appendix A therefore gives, similarly to (3.10),

$$\max_{1\le p\le [nt]} \mathrm{E}(u_p^2 1_{\{|u_p|>\varepsilon/c_{np}\}}) = \begin{cases} o(n^{1-r/2}), & d_x > 0 \\ o(n^{(d_x+1/2)(2-r)}), & d_x < 0. \end{cases} \tag{6.35}$$

These conditions, similarly to (3.10) and (3.11), confirm that the Lindeberg condition holds in both cases, and also for each $t < 1$.

Now consider $q_{2n}^N(t)$, where $s \le 0$ and according to Lemma **5.2** the moving average weights have the form $K(n)^{-1} a_{n[nt][ns]} = A_2(t,s) + o(1)$ where by (5.10) and (5.19), for $s < t$,

$$\begin{aligned} A_2(t,s) = (1-t)^{d_y}&\left((t-s)^{d_x} Z_1^A(t,s) - (-s)^{d_x}\right) \\ &- 1_{\{t<0\}}(-t)^{d_y}\left((t-s)^{d_x} Z_2^A(t,s) - (-s)^{d_x}\right). \end{aligned} \tag{6.36}$$

As s ranges over $[-N, \min\{0,t\}]$ a salient feature of the function Z_1^A is that

$$Z_1^A(t,s) - 1 = d_y \int_0^1 \tau^{d_y-1}\left(\left(\frac{1-t}{t-s}\tau + 1\right)^{d_x} - 1\right) \mathrm{d}\tau = O((-s)^{-1}) \tag{6.37}$$

and $Z_2^A(t,s) = 1 + O((-s)^{-1})$ similarly.

Consider first the case $t \ge 0$. Then (6.36) gives

$$A_2(t,s) = (1-t)^{d_y}\left((t-s)^{d_x} - (-s)^{d_x} + (t-s)^{d_x}(Z_1^A(t,s) - 1)\right) \tag{6.38}$$

and as $s \to -\infty$,

$$(t-s)^{d_x} - (-s)^{d_x} \sim t d_x (-s)^{d_x-1}. \tag{6.39}$$

In view of (6.37), this implies $|A_2(t,s)| = O((-s)^{d_x-1})$. Setting $p = [ns]$ as before, but this time in the range $-nN \le p \le 0$, in view of (6.39) the constant array has the form

$$c_{np} = \frac{|a_{n,[nt],[nt]-p}|}{\sqrt{n}K(n)} \lesssim \frac{|p|^{d_x-1}}{n^{d_x+1/2}} \tag{6.40}$$

where the notation '$\backsimeq$' is here to be interpreted specifically to mean that the approximating expression contains a factor that can be treated as $O(1)$ as $n \to \infty$. The parallel is with (6.33), but the more complicated form of the extra factor from (6.38) is left implicit.

To apply the Lindeberg condition test to these data points, the approach described on page 40 can be adopted. Divide the terms of $q_{2n}^N(t)$ into N blocks of length n observations, labelled $k = 0, \ldots, N-1$ so that $1-n(k+1) \leq p \leq -nk$ in the kth block. At $k = 0$, the maximum of c_{np} over $[0, 1-n]$ is at $p = 0$ whether $d_x > 0$ or $d_x < 0$. However, according to formula (5.5), $c_{n0} = O(n^{-1/2})$ if $d_x > 0$, whereas in the case $d_x < 0$, $c_{n0} = O(n^{-d_x-1/2})$. (Note, $p = 0$ is the point at which b_0 alone is removed from the inner sum and $|a_{n,[nt],[nt]}| \backsimeq 1$.) Therefore, similarly to (6.35),

$$\max_{1-n\leq p\leq 0} \mathrm{E}(u_p^2 1_{\{|u_p|>\varepsilon/c_{np}\}}) = \begin{cases} o(n^{1-r/2}), & d_x > 0 \\ o(n^{(d_x+1/2)(2-r)}), & d_x < 0. \end{cases}$$

For the cases $1 \leq k \leq N-1$, the maximum of c_{np} in each block is found at $p = -nk$ and according to (6.40),

$$c_{n,-nk}^2 \backsimeq \frac{|-nk|^{2d_x-2}}{n^{2d_x+1}} = O(n^{-3}k^{2d_x-2}).$$

Therefore, after simplification Theorem **A.5** gives

$$\begin{aligned} \sum_{p=1-n(k+1)}^{-nk} c_{np}^2 \mathrm{E}(u_p^2 1_{\{|c_{np}u_p|>\varepsilon\}}) &\leq n \max_{1-n(k+1)\leq p\leq -nk} c_{np}^2 \mathrm{E}(u_p^2 1_{\{|u_p|>\varepsilon/c_{np}\}}) \\ &= o(n^{1-3r/2}k^{(d_x-1)r}). \end{aligned}$$

These bounds are of small order in n and also summable over k for $d_x < \frac{1}{2}$. Since the N blocks are independent by assumption their sum $q_{2n}^N(t)$ converges to a Gaussian limit.

In the case $t < 0$ there are two changes to this procedure. The second term of (6.36) is now nonzero. However, by the same arguments as before this has the form

$$\begin{aligned} &(-t)^{d_y}\left((t-s)^{d_x}Z_2^A(t,s) - (-s)^{d_x}\right) \\ &\quad = (-t)^{d_y}\left(((t-s)^{d_x} - (-s)^{d_x}) + (t-s)^{d_x}(Z_2^A(t,s) - 1)\right) \\ &\quad = O((-s)^{d_x-1}) \end{aligned} \tag{6.41}$$

and hence it is still the case that $|A_2(t,s)| = O((-s)^{d_x-1})$. The equivalence in (6.40) is therefore validated for all t. The second change is that the sum of terms is initialized at $p = [nt]$ instead of $p = 0$. Here, the required modifications to the formulae are most simply made by setting $c_{np} = 0$ for $p \geq [nt]$ with the Lindeberg test being done only for the surviving blocks. If this results in a block with an observation count of small order in n, the contribution of the block to the limiting distribution is negligible by the same token.

Since $A_2^2(t,s) = O((-s)^{2d_x-2})$ as $s \to \infty$, the variance $V^N(t)$ in (6.29) is finite in the limit as $N \to \infty$.

Case 2: $d_y < 0$.

Here the decomposition is applied to $A^*(t,s)$ from (5.32). Using the substitutions in (5.35),

$$A^*(t,s) = 1_{\{s>0\}}A_1^*(t,s) + 1_{\{s\leq 0\}}A_2^*(t,s) \tag{6.42}$$

where for $s < t$,

$$A_1^*(t,s) = (1-t)^{d_y}(1-s)^{d_x} - \frac{d_x}{d_y+1}(1-t)^{d_y+1}(t-s)^{d_x-1}Z_1^{A*}(t,s) \tag{6.43}$$

with $Z_1^{A*}(t,s)$ from (5.33) and also, with $Z_2^{A*}(t,s)$ from (5.34),

$$A_2^*(t,s) = (1-t)^{d_y}\left((1-s)^{d_x} - (-s)^{d_x}\right) - \frac{d_x}{d_y+1}(t-s)^{d_x-1} \times \left((1-t)^{d_y+1}Z_1^{A*}(t,s) - 1_{\{t<0\}}(-t)^{d_y+1}Z_2^{A*}(t,s)\right). \tag{6.44}$$

It may be verified as before that $Z_1^{A*}(t,s) = 1 + O((-s)^{-1})$ and $Z_2^{A*}(t,s) = 1 + O((-s)^{-1})$.

Considering the case $q_{1n}(t)$, with $0 < s < t$, the same sequence of arguments can be followed up to (6.33). However, now rewrite (6.43) in the form

$$A_1^*(t,s) = (1-t)^{d_y}(t-s)^{d_x}\left(\left(\frac{1-s}{t-s}\right)^{d_x} - \frac{d_x}{d_y+1}\frac{1-t}{t-s}Z_1^{A*}(t,s)\right). \tag{6.45}$$

As before, set $p = [ns]$ with t fixed and so similarly to the case in (6.33) write

$$c_{np} \backsimeq \frac{([nt]-p)^{d_x}}{n^{d_x+1/2}} \tag{6.46}$$

where the symbol '$\backsimeq$' denotes limiting proportionality with the bracketed term in (6.45). While depending on s, this is $O(1)$ as $s \to 0$. The Lindeberg condition is verified as in (6.34).

For the case $s \le 0$, using the mean value theorem to replace $(1-s)^{d_x} - (-s)^{d_x}$, (6.44) can be rearranged as

$$A_2^*(t,s) = (t-s)^{d_x-1}\Bigg((1-t)^{d_y} d_x \left(\frac{\lambda - s}{t-s}\right)^{d_x-1}$$
$$- \frac{d_x}{d_y+1}\Big(\big((1-t)^{d_y+1} - 1_{\{t<0\}}(-t)^{d_y+1}\big) + (1-t)^{d_y+1}\left(Z_1^{A*}(t,s) - 1\right)$$
$$-1_{\{t<0\}}(-t)^{d_y+1}\left(Z_2^{A*}(t,s) - 1\right)\Big)\Bigg). \tag{6.47}$$

Here, $\lambda \in [0,1]$ depends on s but nonetheless the expression in large parentheses is $O(1)$ as $s \to -\infty$ with t fixed, with the last two terms being of $O((-s)^{-1})$. In this range, therefore,

$$c_{np} \backsimeq \frac{|p|^{d_x-1}}{n^{d_x+1/2}} \tag{6.48}$$

as in the matching result in (6.40). With (6.47) in place of (6.44), the Lindeberg condition can be checked in the same way as for $d_y > 0$.

Since $A_2^{*2}(t,s) = O((-s)^{2d_x-2})$, the limiting variance in (6.29) is similarly finite as $N \to \infty$. ∎

6.4 Almost Sure Continuity

Lemma **6.12** has shown that the partial sum processes $q_n^N(t)$ have Gaussian limits. It remains to show that Q^N, the limit of the process q_n^N composed of the terminal points of these processes, is a.s. continuous.

An increment of q_n^N is the sum of two components, due respectively to the addition of terms and the change in moving average coefficients of existing terms. The decomposition has the form

$$q_n^N(t+\delta) - q_n^N(t) = Y_{1n}(t+\delta, t) + Y_{2n}(t+\delta, t) \tag{6.49}$$

where

$$Y_{1n}(s,t) = \frac{1}{\sqrt{n}K(n)} \sum_{p=[nt]+1}^{[ns]} a_{n[ns]p} u_p \tag{6.50a}$$

$$Y_{2n}(s,t) = \frac{1}{\sqrt{n}K(n)} \sum_{p=1-nN}^{[nt]} (a_{n[ns]p} - a_{n[nt]p}) u_p. \tag{6.50b}$$

Lemma **6.12** shows that (6.49) is Gaussian in the limit under Assumption **4.1**, noting in particular that $Y_{1n}(s,t)$ and $Y_{2n}(s,t)$ are nonoverlapping and independent of each other and that $Y_{2n}(s,t)$ is a sum whose weights meet the conditions tested in the lemma. That is, they are bounded absolutely by those of either $2q_{2n}^N(t)$ or $2q_{2n}^N(s)$ as defined in (6.30).

Also define $v_n^2(t,\delta) = v_{1n}^2(t,\delta) + v_{2n}^2(t,\delta)$ where

$$v_{1n}^2(t,\delta) = \frac{1}{nK(n)^2}\sum_{p=[nt]+1}^{[n(t+\delta)]} a_{n[n(t+\delta)]p}^2 \tag{6.51a}$$

$$v_{2n}^2(t,\delta) = \frac{1}{nK(n)^2}\sum_{p=1-nN}^{[nt]} (a_{n[n(t+\delta)]p} - a_{n[nt]p})^2 \tag{6.51b}$$

and under Assumption **4.1**,

$$\mathrm{E}(q_n^N(t+\delta) - q_n^N(t))^2 = \sigma_u^2 v_n^2(t,\delta).$$

What now needs to be shown, similarly to the developments in §3.3, is that these increments are uniformly bounded in probability and uniformly integrable in the sense of Lemma **3.6**. There are parallels here with Theorem **3.5**, as the following result shows.

6.13 Theorem Under Assumption **4.1**, $\sup_{t\le s\le t+\delta}|Y_{1n}(s,t)|$ is bounded in probability if and only if Assumption **6.9**(a) holds.

Proof According to (5.5),

$$\begin{aligned} Y_{1n}(s,t) = \frac{1}{\sqrt{n}K(n)}\Big(&(c_0b_0 + c_1(b_0+b_1) + c_2(b_0+b_1+b_2)+\cdots)\,u_{[ns]} \\ &+ ((c_0(b_0+b_1) + c_1(b_0+b_1+b_2) \\ &\quad + c_2(b_0+b_1+b_2+b_3)+\cdots)\,u_{[ns]-1}+\cdots\Big). \end{aligned} \tag{6.52}$$

When $d_x + d_y < 0$ the coefficient of $u_{[ns]}$ in (6.52) is $O(n^{-(1/2+d_x+d_y)})$ as $n \to \infty$, since prior to normalization the terms are summable. The argument is identical to that of Theorem **3.5** except that $\kappa(n)$ in (3.20) is replaced with $\sqrt{n}K(n)$. Assumption **6.9**(a) is sufficient for $(\sqrt{n}K(n))^{-r} = O(n^{-1})$. ∎

Similarly to Theorem **3.5** and (3.24) in the remark following, Theorem **6.13** shows that Assumption **6.9**(a) is necessary for uniform integrability of the q_n^N increments. Subject to this requirement, the main argument leading to stochastic equicontinuity is shown, just as in the univariate case, by a combination of Theorem **A.8** of Appendix A and a modified version of Lemma **3.6**. The moving average coefficients are constructed differently here from the univariate case, but the basic idea of finding a uniformly integrable dominating sequence for the supremum is the same.

Adopting the compact notation introduced for (3.18) on page 42, let

$$\widetilde{T}_n^2(t,\delta) = \sup_{\{t\le s\le t+\delta\}} \frac{(q_n^N(s) - q_n^N(t))^2}{v_n^2(t,\delta)} \tag{6.53}$$

and similarly define

$$\widetilde{Y}_{1n}(t,\delta) = \sup_{\{t\le s\le t+\delta\}} \frac{|Y_{1n}(s,t)|}{v_{1n}(t,\delta)}, \quad \widetilde{Y}_{2n}(t,\delta) = \sup_{\{t\le s\le t+\delta\}} \frac{|Y_{2n}(s,t)|}{v_{2n}(t,\delta)}.$$

6.14 Lemma Under Assumptions **4.1** and **6.9**, the collection $\{\widetilde{T}_n^2(t,\delta), n \in \mathbb{N}\}$ in (6.53) is uniformly integrable for any t in $(-N, 1)$ and $0 < \delta < 1 - t$, with

$$\mathrm{E}\left(\widetilde{T}_n^2(t,\delta)1_{\{\widetilde{T}_n^2(t,\delta)>\eta\}}\right) = o(\eta^{2-r}). \tag{6.54}$$

Proof Begin by noting that Assumption **6.9** validates the necessary condition of Theorem **6.13**. Then consider the easily verified inequality

$$\widetilde{T}_n^2(t,\delta) \le \left(\widetilde{Y}_{1n}(t,\delta) + \widetilde{Y}_{2n}(t,\delta)\right)^2. \tag{6.55}$$

The approach is to show that each term in the majorant of (6.55) satisfies the conditions of Theorem **A.8** and then to invoke Theorems **A.7** and **A.6**.

Starting with Y_{1n}, consider a modified version of formula (5.5). For fixed integer $b \in \{[nt],\dots,[n(t+\delta)]\}$ and likewise any $p \in \{[nt],\dots,[n(t+\delta)]\}$, define

$$\bar{a}_{nbp} = \sum_{l=\max\{0,1-b,p-b\}}^{n-1-b} c_l \left(\sum_{j=\max\{0,1-p\}}^{l+b-p} b_j\right). \tag{6.56}$$

The main difference from (5.5) is that p is allowed to exceed b. Defining r by $b = [nr]$, the analysis paralleling Lemma **5.2** gives the result

$$\frac{\bar{a}_{n[nr][ns]}}{K(n)} \to d_y \int_{\max\{0,-r,s-r\}}^{1-r} v^{d_y-1}\left((v+r-s)^{d_x} - 1_{\{s<0\}}(-s)^{d_x}\right) \mathrm{d}v. \tag{6.57}$$

Hence, defining

$$\bar{v}_{1nb}^2(t,\delta) = \frac{1}{nK(n)^2} \sum_{p=[nt]+1}^{[n(t+\delta)]} \bar{a}_{nbp}^2 \tag{6.58}$$

note the implication of (6.57) that $\bar{v}_{1nb}^2(t,\delta)/v_{1n}^2(t,\delta) = O(1)$ as $n \to \infty$. In the partial sum sequence

$$\bar{Y}_{1nb}(s,t) = \frac{1}{\sqrt{n}K(n)} \sum_{p=[nt]+1}^{[ns]} \bar{a}_{nbp} u_p \tag{6.59}$$

the moving average weights do not depend on s. Similarly to (6.53) define

$$\tilde{\bar{Y}}_{1nb}^2(t,\delta) = \sup_{\{t \le s \le t+\delta\}} \frac{\bar{Y}_{1nb}^2(s,t)}{v_{1n}^2(t,\delta)} \tag{6.60}$$

taking care to note that in this instance the denominator is chosen to be $v_{1n}^2(t,\delta)$, not $\bar{v}_{1nb}^2(t,\delta)$. Since their ratio is $O(1)$, the conditions of Theorem **A.8** are met with either choice of divisor. For any choice of b in the indicated range, the collection $\left\{\tilde{\bar{Y}}_{1nb}^2(t,\delta), n \in \mathbb{N}\right\}$ is uniformly integrable and satisfies the relation

$$\mathrm{E}\left(\tilde{\bar{Y}}_{1nb}^2(t,\delta) 1_{\{\tilde{\bar{Y}}_{1nb}(t,\delta) > \eta\}}\right) = o(\eta^{2-r}).$$

Now set $b = [ns^*]$ where s^* is the solution to $\sup_{\{t \le s \le t+\delta\}} Y_{1n}^2(s,t)/v_{1n}^2(t,\delta)$, so that

$$\sup_{\{t \le s \le t+\delta\}} \frac{Y_{1n}^2(s,t)}{v_{1n}^2(t,\delta)} = \frac{\bar{Y}_{1n[ns^*]}^2(s^*,t)}{v_{1n}^2(t,\delta)} \le \sup_{\{t \le s \le t+\delta\}} \frac{\bar{Y}_{1n[ns^*]}^2(s,t)}{v_{1n}^2(t,\delta)}. \tag{6.61}$$

The collection of the majorants of (6.61) for each $n \in \mathbb{N}$ is uniformly integrable, since each member is drawn from one of the uniformly integrable collections defined for (6.60). The result that the collection $\left\{\tilde{Y}_{1n}^2(t,\delta), n \in \mathbb{N}\right\}$ is uniformly integrable, with the property $\mathrm{E}\left(\tilde{Y}_{1n}^2(t,\delta) 1_{\{\tilde{Y}_{1n}(t,\delta) > \eta\}}\right) = o(\eta^{2-r})$ for each $n \in \mathbb{N}$, follows by Theorem **A.6**.

The second term of (6.55) is not a partial sum and depends on s only through the definition of the moving average weights. For this sum, argue as follows. Since $v_{2n}^2(t, s-t) \le v_{2n}^2(t,\delta)$ for $s \le t+\delta$ when n is large enough, by a minor extension of Theorem **6.5**(ii), it can be verified from (6.50b) and (6.51b) that the collection $\{Y_{2n}^2(t,s)/v_{2n}^2(t,\delta), n \in \mathbb{N}\}$ satisfies the conditions of Corollary **A.9**. This is true for each $s \in [t, t+\delta]$ and hence in particular, it is true for $\widetilde{Y}_{2n}(t,\delta)$. The theorem now follows by application of Theorems **A.7** and **A.6** to (6.55). ∎

At this point, it is possible to set out the formal proof of the weak convergence result from §6.2 for the process $q_n^N = \{q_n^N(t), t \in [-N, 1]\}$.

Proof of Theorem 6.11 Given Lemmas **6.12** and **6.14**, the proof for q_n^N aligns closely with that of Theorem **3.2**. To show stochastic equicontinuity, apply Theorems **3.7** and **3.8** with Y_n in (3.35) equated to q_n^N with the sum of (6.51a) and (6.51b) playing the role of $v_n^2(t,\delta)$. In this application of the theorems, $L = -N$ and $U = 1$ and also $d = d_x$. Condition (a) of Theorem **3.7** is satisfied according to Theorem **6.13**. Condition (b) of Theorem **3.7** is shown by confirming the conditions of Theorem **3.8**. Since $r \geq 1/(\frac{1}{2} + \min\{0, d_x, d_y + d_x\})$ by Assumption **6.9**(a), the exponent of δ in condition (3.41) with the substitution $d = d_x$ is nonnegative at worst, confirming that Theorem **3.8** holds in this case subject to confirmation of the stated conditions. Of these, condition **3.8**(b) is shown in Lemma **6.14** while condition **3.8**(a) holds by Theorem **6.5**. ∎

The parallel result for h_n^N follows closely similar lines and is stated formally as follows.

6.15 Theorem Under Assumptions **4.1** and **6.9**(b), $h_n^N \to_d H^N$ where the limit is an a.s. continuous Gaussian process on the interval $(-N, 1)$. □

The proof of Theorem **6.15** differs from that of Theorem **6.11** only because the functional forms of e_{npm} and hence of $E(s,t)$ and $E^*(s,t)$ replace those of a_{nmp}, $A(t,s)$ and $A^*(t,s)$. Making such substitutions as

$$c_{nm} = \frac{|e_{n,[ns],[ns]-m+1}|}{\sqrt{n}K(n)} \sim (1-s)^{d_x} Z_1^E(s,t) \frac{([ns]-m)^{d_y}}{n^{d_y+1/2}}$$

to replace (6.33), for example, the convergence criteria are derived in a wholly similar manner. The stochastic equicontinuity also follows by the argument paralleling Lemma **6.14**. Letting h_n^N replace q_n^N in (6.53) the same notation and text can be recycled, with the only significant variations being the replacement of $\bar{a}_{nbp}$ in (6.56) by

$$\bar{e}_{nbm} = \sum_{j=\max\{0,1-b,m-b\}}^{n-1-b} b_j \left(\sum_{l=j+b+1-m}^{n-m} c_l \right)$$

and the citation of Theorem **6.6** in place of Theorem **6.5**. The variance function, in parallel with (6.29), has the form under Assumption **4.1** of

$$\mathrm{E}(H^N(s)^2) = \begin{cases} \sigma_w^2 \int_{-N}^{s} E^2(s,t)\mathrm{d}t & d_x > 0 \\ \sigma_w^2 \int_{-N}^{s} E^{*2}(s,t)\mathrm{d}t & d_x < 0. \end{cases} \tag{6.62}$$

6.5 Stochastic Integral Convergence

It remains to show the convergence of the last two elements of (6.27). The following argument applies almost identically to G_{1n}^N and G_{3n}^N, with G_{1n}^N being taken as usual as the exemplar case.

6.16 Theorem Under Assumptions **4.1** and **6.9**(a), $G_{1n}^N \to_{\mathrm{d}} \Xi_1^N$.

Proof From (6.5) and (6.6),

$$G_{1n}^N = \frac{1}{nK(n)} \sum_{m=-nN}^{n-1} \sum_{p=-nN}^{m} a_{nmp} u_p w_{m+1}. \tag{6.63}$$

Choose an integer subsequence k_n, with $k_n \to \infty$ but $k_n/n \to 0$ as $n \to \infty$. Then, for $j = 0, \ldots, k_n$ let $n_j = [n(N+1)j/k_n] - nN$ so that $n_0 = -nN$, $n_{k_n} = n$, and $n_j - n_{j-1} \to \infty$ for every $j \geq 1$. Finally, define $t_j = n_j/n$ so that $\{t_0, \ldots, t_{k_n}\}$ is a partition of $[-N, 1]$ with the property

$$\max_{1 \leq j \leq k_n} (t_j - t_{j-1}) = O(1/k_n) \tag{6.64}$$

as $n \to \infty$.

With this notation, define a partially aggregated version of the covariance function,

$$\begin{aligned} G_{1n}^{N*} &= \sum_{j=1}^{k_n} q_n^N(t_{j-1}) \left(w_n^N(t_j) - w_n^N(t_{j-1}) \right) \\ &= \frac{1}{nK(n)} \sum_{j=1}^{k_n} \left(\sum_{p=n_0}^{n_{j-1}} a_{nn_{j-1}p} u_p \sum_{m=n_{j-1}+1}^{n_j} w_m \right). \end{aligned} \tag{6.65}$$

Consider the normalized increments in (6.65), which in the case of q_n^N are for each j the partial sums of the elements $q_{n,-nN}^N, \ldots, q_{n,n_{j-1}}^N$ as defined in (6.6). On the assumptions, both the inner sums in (6.65) converge to a.s. continuous Gaussian limits, respectively Q^N by Theorem **6.11** and W^N by Theorem **3.2** for the case $d = 0$. Replacing q_n^N and w_n^N in (6.65) by these limit processes, define

$$P_n = \sum_{j=1}^{k_n} Q^N(t_{j-1})\left(W^N(t_j) - W^N(t_{j-1})\right) = \sum_{j=1}^{k_n} Q^N(t_{j-1}) \int_{t_{j-1}}^{t_j} \mathrm{d}W^N(t). \tag{6.66}$$

Take care to note that although P_n depends upon n it is a functional of the limit processes Q^N and W^N.

The tasks of the proof are to show the connections between G_{1n}^N and G_{1n}^{N*}, P_n and Ξ_1^N, and finally between G_{1n}^{N*} and P_n. The sequences of the differences between G_{1n}^N and G_{1n}^{N*} and P_n and Ξ_1^N are shown to converge to zero in mean square as $n \to \infty$. It is also shown by an indirect method that G_{1n}^{N*} and P_n have the same limit in distribution.

The first of these remainders has the form

$$\begin{aligned} G_{1n}^N - G_{1n}^{N*} = \frac{1}{nK(n)} \sum_{j=1}^{k_n} \sum_{m=n_{j-1}+1}^{n_j-1} \Bigg(\sum_{p=n_{j-1}+1}^{m} & a_{nmp} u_p \\ &+ \sum_{p=n_0}^{n_{j-1}} (a_{nmp} - a_{nn_{j-1}p}) u_p \Bigg) w_{m+1}. \end{aligned} \tag{6.67}$$

Since the shocks are L_2-bounded and serially independent and nowhere overlap,

$$\begin{aligned} \mathrm{E}(G_{1n}^N - G_{1n}^{N*})^2 = \frac{\sigma_u^2 \sigma_w^2}{n^2 K(n)^2} \sum_{j=1}^{k_n} \sum_{m=n_{j-1}+1}^{n_j-1} \Bigg(\sum_{p=n_{j-1}+1}^{m} & a_{nmp}^2 \\ &+ \sum_{p=n_0}^{n_{j-1}} (a_{nmp} - a_{nn_{j-1}p})^2 \Bigg). \end{aligned} \tag{6.68}$$

Consider the two sums of squares in (6.68). By Theorem **6.5**(i) and the fact that $n_j - n_{j-1} = O(n/k_n)$,

$$\begin{aligned} \frac{1}{n^2 K(n)^2} \sum_{j=1}^{k_n} \sum_{m=n_{j-1}+1}^{n_j-1} \sum_{p=n_{j-1}+1}^{m} a_{nmp}^2 & \\ &\ll \frac{1}{n} \sum_{j=1}^{k_n} \sum_{m=n_{j-1}+1}^{n_j-1} \left(\frac{m - n_{j-1}}{n} \right)^{\min\{1,1+2d_x\}} \\ &\ll \frac{1}{n^{\min\{2,2+2d_x\}}} \sum_{j=1}^{k_n} (n_j - n_{j-1})^{\min\{2,2+2d_x\}} \\ &= O\left(k_n^{-\min\{1,1+2d_x\}}\right). \end{aligned}$$

For the second block similarly, by Theorem **6.5**(ii) and (6.64),

$$\frac{1}{n^2K(n)^2}\sum_{j=2}^{k_n}\sum_{m=n_{j-1}+1}^{n_j-1}\sum_{p=n_0}^{n_{j-1}}(a_{nmp}-a_{nn_{j-1}p})^2$$

$$\ll \frac{1}{n}\sum_{j=2}^{k_n}\sum_{m=n_{j-1}+1}^{n_j-1}\left(\frac{m-n_{j-1}}{n}\right)^{1+2d_x}$$

$$= O\left(k_n^{-1-2d_x}\right).$$

In other words,

$$\mathrm{E}(G_{1n}^N - G_{1n}^{N*})^2 = O\left(\max\{k_n^{-1}, k_n^{-1-2d_x}\}\right) \tag{6.69}$$

which with $d_x > -\frac{1}{2}$ implies $|G_{1n}^N - G_{1n}^{N*}| \to_{L_2} 0$ in all cases.

Next, compare P_n in (6.66) with $\Xi_1^N = \int_{-N}^1 Q^N(t)\mathrm{d}W^N(t)$. The fact that $\mathrm{E}\left(Q^N(t+\delta) - Q^N(t)\right)^2 = O(\delta^{2d_x+1})$ can be deduced from the formula in (5.26) combined with Lemma **6.1** and the calculations in Theorem **6.5**. Using this bound and the fact that W^N is a Brownian motion gives

$$\mathrm{E}\left(P_n - \Xi_1^N\right)^2 = \mathrm{E}\left(\sum_{j=1}^{k_n}\int_{t_{j-1}}^{t_j}\left(Q^N(t_{j-1}) - Q^N(t)\right)\mathrm{d}W^N(t)\right)^2$$

$$\ll \sum_{j=1}^{k_n}\int_{t_{j-1}}^{t_j}(t-t_{j-1})^{2d_x+1}\mathrm{d}t$$

$$\ll \sum_{j=1}^{k_n}(t_j - t_{j-1})^{2d_x+2} = O(k_n^{-2d_x-1}). \tag{6.70}$$

The final order of magnitude is by (6.64). In other words, $|P_n - \Xi_1^N| \to_{L_2} 0$.

The remaining step is to connect G_{1n}^{N*} with P_n in (6.66), and this is done by applying the Skorokhod representation theorem (see [71]),[1] according to which the joint weak convergence of q_n^N and w_n^N to Q^N and W^N implies the existence of processes that are distributed like q_n^N and w_n^N and converge almost surely to limits distributed like Q^N and W^N. These so-called Skorokhod processes will be denoted $\hat{q}_n^N$ and $\hat{w}_n^N$. If $\widehat{G}_{1n}^{N*}$ is the counterpart of G_{1n}^{N*} evaluated with $\hat{q}_n^N$ and $\hat{w}_n^N$ in place of q_n^N and w_n^N then $\widehat{G}_{1n}^{N*}$ and G_{1n}^{N*} have the same distribution. To show that G_{1n}^{N*} is distributed like P_n when n is large enough it suffices to show that $|\widehat{G}_{1n}^{N*} - P_n| \to_{L_2} 0$.

[1] Proved as SLT Theorem 29.6.

The first step in this demonstration is by way of the easily verifiable identity

$$
\begin{aligned}
&\widehat{G}_{1n}^{N*} - P_n \\
&= \sum_{j=1}^{k_n} \widehat{q}_n^N(t_{j-1})\left(\widehat{w}_n^N(t_j) - \widehat{w}_n^N(t_{j-1})\right) - \sum_{j=1}^{k_n} Q^N(t_{j-1})\left(W^N(t_j) - W^N(t_{j-1})\right) \\
&= \sum_{j=1}^{k_n} \left(\widehat{q}_n^N(t_{j-1}) - Q^N(t_{j-1})\right)\left(\widehat{w}_n^N(t_j) - \widehat{w}_n^N(t_{j-1})\right) \\
&\qquad - \sum_{j=1}^{k_n} \left(Q^N(t_j) - Q^N(t_{j-1})\right)\left(\widehat{w}_n^N(t_j) - W^N(t_j)\right) \\
&\quad + Q^N(1)\left(\widehat{w}_n^N(1) - W^N(1)\right) - Q^N(-N)\left(\widehat{w}_n^N(-N) - W^N(-N)\right). \qquad (6.71)
\end{aligned}
$$

It is sufficient to show that each of the four right-hand side terms of (6.71) converges to zero in mean square. Starting with the first one, the Cauchy–Schwarz inequality for sums gives

$$
\begin{aligned}
&\left(\sum_{j=1}^{k_n} \left(\widehat{q}_n^N(t_{j-1}) - Q^N(t_{j-1})\right)\left(\widehat{w}_n^N(t_j) - \widehat{w}_n^N(t_{j-1})\right)\right)^2 \\
&\quad \le \sum_{j=1}^{k_n} \left(\widehat{q}_n^N(t_{j-1}) - Q^N(t_{j-1})\right)^2 \sum_{j=1}^{k_n} \left(\widehat{w}_n^N(t_j) - \widehat{w}_n^N(t_{j-1})\right)^2 \\
&\quad \le k_n \max_{1\le j\le k_n} \left(\widehat{q}_n^N(t_{j-1}) - Q^N(t_{j-1})\right)^2 \sum_{j=1}^{k_n} \left(\widehat{w}_n^N(t_j) - \widehat{w}_n^N(t_{j-1})\right)^2. \qquad (6.72)
\end{aligned}
$$

The object is to bound the expected value of the majorant of (6.72).

The process $\widehat{q}_n^N$ in (6.72) is càdlàg, a step function with discontinuities (see page 23 for details), although its almost sure limit Q^N is a.s. continuous according to Theorem **6.11**. A gentle digression into probability theory is necessary at this point. Let $(\Omega, \mathcal{F}, P)$ be the probability space where these random processes reside and let $\omega \in \Omega$ denote an outcome. By Egorov's theorem,[2] almost sure convergence implies uniform convergence on a set of outcomes $A_\varepsilon \in \mathcal{F}$, where $P(A_\varepsilon) \ge 1 - \varepsilon$ and $\varepsilon > 0$ is arbitrary. For càdlàg functions, which inhabit a space of processes endowed with the Skorokhod topology, what this amounts to is that if $\widehat{q}_n^N(\omega) \to Q^N(\omega)$ almost surely then,

[2] See SLT Theorem 19.4.

for A_ε so defined,

$$\sup_{\omega \in A_\varepsilon} d_S\left(\widehat{q}_n^N(\omega), Q^N(\omega)\right) \to 0$$

where d_S is the Skorokhod distance discussed on page 23. Since Q^N is a.s. continuous, there also exists $E_Q \in \mathcal{F}$ with $P(E_Q) = 1$ such that each $\omega \in E_Q$ has the following property: for any $\eta > 0$, there exists a constant $\delta > 0$ such that if

$$\sup_{t \in [-N,1]} \left|\widehat{q}_n^N(\omega, t) - Q^N(\omega, \lambda(t))\right| \leq \delta$$

where $\lambda : [-N, 1] \mapsto [-N, 1]$ represents the homeomorphism defining the Skorokhod distance, then

$$\begin{aligned}
&\sup_{t \in [-N,1]} \left|\widehat{q}_n^N(\omega, t) - Q^N(\omega, t)\right| \\
&\quad \leq \sup_{t \in [-N,1]} \left|\widehat{q}_n^N(\omega, t) - Q^N(\omega, \lambda(t))\right| + \sup_{t \in [-N,1]} \left|Q^N(\omega, \lambda(t)) - Q^N(\omega, t)\right| \\
&\quad \leq \delta + \eta.
\end{aligned}$$

In other words, this says that when a càdlàg function is close to a continuous function the Skorokhod distance must be correspondingly close to the uniform distance. Letting $A_\varepsilon^* = A_\varepsilon \cap E_Q$ so that $P(A_\varepsilon^*) = P(A_\varepsilon)$, it follows that $\delta_n^q \to 0$ as $n \to \infty$ where

$$\delta_n^q = \sup_{\omega \in A_\varepsilon^*} \sup_{t \in [-N,1]} \left|\widehat{q}_n^N(\omega, t) - Q^N(\omega, t)\right|.$$

Noting that the distributions of w_n^N and $\widehat{w}_n^N$ are the same, (6.72) implies that

$$\begin{aligned}
&\mathrm{E}\left(1_{A_\varepsilon^*} \sum_{j=1}^{k_n} \left(\widehat{q}_n^N(t_{j-1}) - Q^N(t_{j-1})\right)\left(\widehat{w}_n^N(t_j) - \widehat{w}_n^N(t_{j-1})\right)\right)^2 \\
&\qquad \leq k_n(\delta_n^q)^2 \sum_{j=1}^{k_n} \mathrm{E}\left(w_n^N(t_j) - w_n^N(t_{j-1})\right)^2 \\
&\qquad = \sigma_w^2 k_n (\delta_n^q)^2 \sum_{j=1}^{k_n} (t_j - t_{j-1})^2 = O((\delta_n^q)^2)
\end{aligned}$$

where the final equality is by (6.64). Letting $\varepsilon \to 0$ and hence $P(A_\varepsilon^*) \to 1$, it follows that the first term of (6.71) converges to zero in mean square.

The same type of argument may be applied to the second term of (6.71). $\widehat{w}_n^N$ is also a càdlàg process converging with probability 1 to an a.s. continuous limit, and there exists a set A_ε^* with $P(A_\varepsilon^*) \geq 1 - \varepsilon$ such if

$$\delta_n^w = \sup_{\omega \in A_\varepsilon^*} \sup_{t \in [-N,1]} \left| \widehat{w}_n^N(\omega, t) - W^N(\omega, t) \right|$$

then $\delta_n^w \to 0$ as $n \to \infty$. It follows in this case that

$$\begin{aligned} \mathrm{E}\left(1_{A_\varepsilon^*} \sum_{j=1}^{k_n} (Q^N(t_j) - Q^N(t_{j-1})) \left(\widehat{w}_n^N(t_j) - W^N(t_j) \right) \right)^2 & \\ \leq k_n (\delta_n^w)^2 \sum_{j=1}^{k_n} \mathrm{E} \left(Q^N(t_j) - Q^N(t_{j-1}) \right)^2 & \\ = O(k_n^{-2d_x} (\delta_n^w)^2) & \end{aligned}$$

where the order of magnitude is by reasoning similar to (6.70). When $d_x < 0$ it is necessary that k_n diverge slowly enough that $k_n^{-2d_x}(\delta_n^w)^2 \to 0$, which since $d_x > -\frac{1}{2}$ is possible under the assumptions. The expected squares of the third and fourth terms of (6.71) are of $O((\delta_n^w)^2)$ and so vanish under the same assumptions. ∎

Under Assumption **4.1**, it appears reasonable to conjecture that $\delta_n^w = O(n^{-1/2})$. If that were the case, any $k_n = o(n)$ would satisfy the indicated convergence.

The counterpart result for G_{3n}^N, with only minor variations in the argument of Theorem **6.16**, is stated for completeness.

6.17 Theorem Under Assumptions **4.1** and **6.9**(b), $G_{3n}^N \to_{\mathrm{d}} \Xi_3^N$. □

It remains to gather together the various results of this chapter to address the proposition initially posed.

Proof of Theorem 6.10 The different elements of (6.27) are accounted for, in the first instance, by Theorem **3.2** applied with $d = 0$ to give regular Brownian motion limits to both w_n^N and u_n^N. Theorems **6.11**, **6.15**, **6.16**, and **6.17** yield the remaining elementwise limits. The six elements are all adapted to the same filtration and by defining functions equated everywhere to the fixed values G_{1n}^N and G_{3n}^N they can be embedded in $D_{[-N,1]}^6$, converging to a limit lying in $C_{[-N,1]}^6$ almost surely. Theorem **3.19** delivers the required joint convergence. ∎

Chapter 7
Fractional Cointegration

7.1 Stationary Regression

This chapter reviews some implications of the foregoing analysis for the interpretation of linear regressions. To begin with, consider the regression model

$$p_i = \alpha + \beta x_i + y_i \tag{7.1}$$

where x_i and y_i are stationary zero-mean processes, α and β are parameters, and p_i is defined by the equation. The assumption to be maintained is that either or both of x_i and y_i are fractional processes, and that Assumption **4.1** applies. In this chapter, slowly varying components in (1.2) are omitted for the sake of simplicity. While these are dealt with in practice by modifying normalizations so that they disappear from limit expressions, the important fact to note is that the conditions encountered on the behaviour of limit distributions are mainly expressed as strict inequalities on functions of d_x and d_y. In these cases, slowly varying factors would not affect the result whether or not they are included in the normalization.

The OLS error-of-estimate for the slope coefficient β in (7.1), after the conventional normalization by $\sqrt{n}$, has the well-known formula

$$\sqrt{n}(\hat{\beta} - \beta) = \frac{n^{-1/2} \sum_i x_i y_i - n^{-3/2} \sum_i x_i \sum_i y_i}{n^{-1} \sum_i x_i^2 - \left(n^{-1} \sum_i x_i\right)^2}. \tag{7.2}$$

Consider what can be said about each of the terms in (7.2). If u and w are independent and $d_x + d_y < \frac{1}{2}$, the first term of the numerator is asymptotically Gaussian by Theorem **4.3**, with variance $V_{xy} < \infty$ from (4.9). In the second term of the numerator, the two sums are respectively $O_p(n^{1/2+d_x})$ and $O_p(n^{1/2+d_y})$ by Corollary **2.7**, and if $1 + d_x + d_y < \frac{3}{2}$ this term is of small order relative to the first. In the denominator, Theorem **4.2** gives $n^{-1} \sum_i x_i^2 \to_{L_2} \sigma_x^2$ with the formula in (1.5), whereas $n^{-1} \sum_i x_i = o_p(1)$ by Corollary **2.7**, so here too the second term is of relatively small order. The conclusion is therefore that

Asymptotics for Fractional Processes. James Davidson, Oxford University Press. © James Davidson (2025).
DOI: 10.1093/9780198955207.003.0007

$$\sqrt{n}(\hat{\beta} - \beta) \xrightarrow{d} \mathrm{N}\left(0, \frac{V_{xy}}{\sigma_x^4}\right). \tag{7.3}$$

In the case with $d_y = 0$ so that $y_i = w_i$ with variance σ_w^2, the remark following the proof of Theorem **4.3** on page 73 shows that $V_{xy} = \sigma_w^2\sigma_x^2$ and the variance of $\sqrt{n}(\hat{\beta} - \beta)$ has the usual limit of σ_w^2/σ_x^2. In this case, x is merely required to be stationary. Further discussion of the restriction $d_x + d_y < \frac{1}{2}$ can be found in §7.3.

Although the regressor in (7.1) has been assumed to have a zero mean, this involves no loss of generality. Recall that a stationary fractional process can have a mean $\mu \neq 0$ only in the form $x_i + \mu$ where x_i has the specification of (1.1)+(1.2). Since the least squares formula expresses the data in the form of sample mean deviations, it is a simple exercise to show that the limit in (7.3) continues to hold. The terms that would contain μ in (7.2) all cancel with one another in the limit.

If $\sigma_{uw} \neq 0$ on the other hand, the estimator is biased and inconsistent with

$$\hat{\beta} - \beta \xrightarrow{L_2} \frac{\sigma_{xy}}{\sigma_x^2}. \tag{7.4}$$

Also consider the intercept, where several different scenarios are possible since in this case the presence of a regressor mean μ does make a difference. If y is weakly dependent with $d_y = 0$, the normalized error-of-estimate formula is

$$\sqrt{n}(\hat{\alpha} - \alpha) = \frac{n^{-1}\sum_i x_i^2 n^{-1/2}\sum_i y_i - n^{-1}\sum_i x_i n^{-1/2}\sum_i x_i y_i}{n^{-1}\sum_i x_i^2 - \left(n^{-1}\sum_i x_i\right)^2}. \tag{7.5}$$

If x_i is replaced by $x_i + \mu$ in (7.5), a simple calculation gives

$$\sqrt{n}(\hat{\alpha} - \alpha) \sim \frac{1}{\sqrt{n}}\sum_{i=1}^{n} y_i - \mu\sqrt{n}(\hat{\beta} - \beta) \tag{7.6}$$

which, whether $\mu = 0$ or otherwise, has the same normalization as in the weakly dependent regressor case. However, if y is a long memory fractional process with $d_y > 0$ the required normalizing factor becomes $n^{1/2-d_y}$. The term containing μ, if present, is now of small order and by Lemmas **3.3** and **3.4**,

$$n^{1/2-d_y}(\hat{\alpha} - \alpha) \sim \frac{1}{n^{1/2+d_y}}\sum_{i=1}^{n} y_i \xrightarrow{d} \mathrm{N}\left(0, \sigma_w^2\Upsilon_{d_y}\right). \tag{7.7}$$

If $d_y < 0$ on the other hand, the equivalence in (7.7) continues to hold if $\mu = 0$ but otherwise the second term in (7.6) dominates.

7.2 Cointegrating Regression

The leading applications of regression to fractional processes involve cointegrating relations between processes featuring stochastic trends. Long memory and nonstationarity are closely connected in time series analysis and while nonstationarity is conventionally due to a unit root, this becomes a special case in the present analysis. The important questions to be investigated are sufficiently answered by consideration of the bivariate model

$$P_i = \alpha + \beta S_i + y_i \tag{7.8}$$

where now the regressor is the partial sum process $S_i = \sum_{j=1}^{i} x_j$, and x and y are stationary fractional processes with fractional parameters d_x and d_y. P_i is defined by (7.8) so that the model embodies cointegration of nonstationary processes P and S with cointegrating parameter β, although the cointegrating residual can have long memory. Unlike the stationary case, it must be assumed for present purposes that $\mu = 0$ so that there is no drift in the stochastic trend. The relaxation of this restriction is treated in §7.4.

Write the OLS error-of-estimate formula for (7.8), similarly to (7.2) but this time without normalization, as

$$\hat{\beta} - \beta = \frac{n \sum_i S_i y_i - \sum_i S_i \sum_i y_i}{n \sum_i S_i^2 - \left(\sum_i S_i\right)^2}. \tag{7.9}$$

The data moments involved in this calculation, together with the weak limit of each under suitable normalization, are as follows. Define fBm processes X and Y, having the form of (2.1). Then, under Assumption **4.1**, Theorem **3.2** and the continuous mapping theorem give

$$\frac{1}{n^{d_x+3/2}} \sum_{i=1}^{n} S_i \xrightarrow{d} \int_0^1 X(\xi)\mathrm{d}\xi \tag{7.10a}$$

$$\frac{1}{n^{2d_x+2}} \sum_{i=1}^{n} S_i^2 \xrightarrow{d} \int_0^1 X^2(\xi)\mathrm{d}\xi \tag{7.10b}$$

$$\frac{1}{n^{d_y+1/2}} \sum_{i=1}^{n} y_i \xrightarrow{d} Y(1). \tag{7.10c}$$

Similarly, by the various steps leading to (6.1) in Chapter 6 as well as Theorems **4.7** and **4.8** (see pages 75, 82, and 101 for the various symbol definitions)

$$\frac{1}{n^{d_x+d_y+1}}\sum_{i=1}^{n} S_i y_i \xrightarrow{d} \Xi_{xy} + \sigma_{uw}\lambda_{xy} \text{ if } d_x + d_y > 0 \tag{7.11a}$$

$$\frac{1}{n^{d_x+d_y+1}}\sum_{i=1}^{n} S_i y_i \xrightarrow{d} \Xi_{xy} \text{ if } d_x + d_y > -\tfrac{1}{2} \text{ and } \sigma_{uw} = 0 \tag{7.11b}$$

$$\frac{1}{n\log n}\sum_{i=1}^{n} S_i y_i \xrightarrow{L_2} \gamma_{xy} \text{ if } d_x = -d_y \text{ and } \sigma_{uw} \neq 0 \tag{7.11c}$$

$$\frac{1}{n}\sum_{i=1}^{n} S_i y_i \xrightarrow{L_2} \gamma_{xy} + \sigma_{xy} \text{ if } d_x + d_y < 0 \text{ and } \sigma_{uw} \neq 0. \tag{7.11d}$$

The restriction $d_x + d_y > 0$ in (7.11a) is imposed by the conditions of Theorem **4.4** while the restriction $d_x + d_y > -\frac{1}{2}$ in (7.11b) is due to Assumption **6.9**. To interpret these limits, note that $\sum_{i=1}^{n} S_i y_i = nK(n)G_n + \sum_{i=1}^{n} x_i y_i$ according to (4.10), where the sum on the right-hand side is $O_p(n)$ by Theorem **4.2**. If $d_x + d_y > 0$ so that $K(n)$ diverges, the latter term is of small order so that these limits match the limit of G_n. In case (7.11b), $\sigma_{xy} = 0$ and hence this term is of small order even with $d_x + d_y < 0$. Theorem **4.7** provides the rationale for (7.11c) and (7.11d). In (7.11c), according to Theorem **4.7**(i), the expectation of the sum dominates the mean deviation by the factor $\log n$ so that the limit is a constant. The definition of γ_{xy} in this case is (4.43). In case (7.11d) the limit is again a constant, but this time γ_{xy} has the definition in (4.41) and by Theorem **4.7**(ii) the correct normalization is $1/n$ for both parts of the sum, so σ_{xy} appears in the limit in this case.

The second term in the numerator of (7.9) after normalization has the limit

$$\frac{1}{n^{d_x+d_y+2}}\sum_{i=1}^{n} S_i \sum_{j=1}^{n} y_j \xrightarrow{d} \int_0^1 X(\xi)\mathrm{d}\xi\, Y(1).$$

The mean of this random variable is also of interest and might be written as $\sigma_{uw}\kappa_{xy}$. To calculate κ_{xy}, let the companion function for y corresponding to $a_{ni}(s,t)$ in (2.26) for x be defined as

$$g_{nj}(s,t) = \sum_{l=\max\{0,[nt]+1-j\}}^{[ns]-j} c_l.$$

Then, under Assumption **4.1**,

$$\frac{1}{n^{d_x+d_y+2}}\mathrm{E}\left(\sum_{i=1}^{n} S_i \sum_{j=1}^{n} y_j\right)$$

$$= \frac{1}{n^{d_x+d_y+2}}\mathrm{E}\left(\sum_{i=1}^{n}\sum_{k=-\infty}^{i} a_{nk}(i/n,0)u_k \sum_{j=-\infty}^{n} g_{nj}(1,0)w_j\right)$$

$$= \frac{\sigma_{uw}}{n^{d_x+d_y+2}}\sum_{i=1}^{n}\sum_{k=-\infty}^{i} a_{nk}(i/n,0)g_{nk}(1,0). \tag{7.12}$$

By similar reasoning to Theorem **4.4**, (7.12) converges to $\sigma_{uw}\kappa_{xy}$ where

$$\kappa_{xy} = \int_0^1 \left(\int_0^t (t-\xi)^{d_x}(1-\xi)^{d_y}\mathrm{d}\xi + \int_{-\infty}^0 \left((t-\xi)^{d_x} - (-\xi)^{d_x}\right)\left((1-\xi)^{d_y} - (-\xi)^{d_y}\right)\mathrm{d}\xi\right)\mathrm{d}t.$$

Putting these limits together allows the error of estimate distributions in the four cases of (7.11) to be calculated. In cases (7.11a) and (7.11b), writing $\int_0^1 X\mathrm{d}Y = \Xi_{xy} + \sigma_{uw}\lambda_{xy}$, the limiting distribution has the form

$$n^{1+d_x-d_y}(\hat{\beta}-\beta) \xrightarrow{\mathrm{d}} \frac{\int_0^1 X\mathrm{d}Y - \int_0^1 X\mathrm{d}\xi Y(1)}{\int_0^1 X^2\mathrm{d}\xi - \left(\int_0^1 X\mathrm{d}\xi\right)^2}. \tag{7.13}$$

The estimator is consistent provided $d_y < 1 + d_x$ and this restriction holds for all cases under Assumption **4.1**. Stationarity of y is a sufficient condition for cointegration.

If the regressor is endogenous with $\sigma_{uw} \neq 0$ there is asymptotic bias, which in the unit root case is well known (see e.g., [3]) to play a decisive role in finite samples. In case (7.11a) the mean of the numerator in (7.13) is $\sigma_{uw}(\lambda_{xy} - \kappa_{xy})$. Notwithstanding that the shocks are assumed serially independent under Assumption **4.1**, this asymptotic bias is not attributable to the regressor and regressand being dated contemporaneously. The contribution of the contemporaneous x_i to the form of λ_{xy} is negligible, as pointed out on page 77.

In case (7.11b) there is no bias by construction. In case (7.11c), with appropriate normalization the limit distribution has the form

$$\frac{n^{1+2d_x}}{\log n}(\hat{\beta}-\beta) \xrightarrow{\mathrm{d}} \frac{\gamma_{xy}}{\int_0^1 X^2\mathrm{d}\xi - \left(\int_0^1 X\mathrm{d}\xi\right)^2}. \tag{7.14}$$

In case (7.11d), with appropriate normalization the contemporaneous covariance does not vanish and the limit distribution is

$$n^{1+2d_x}(\hat{\beta}-\beta) \xrightarrow{d} \frac{\gamma_{xy}+\sigma_{xy}}{\int_0^1 X^2 d\xi - \left(\int_0^1 X d\xi\right)^2}. \tag{7.15}$$

Since neither of these limits depends on Ξ_{xy}, (7.14) and (7.15) hold even under the baseline condition of Assumption **4.1**(a), without the restriction on $d_x + d_y$ appearing in (7.11b).

Similarly to (7.5) but without normalization, the error of estimate for the intercept is

$$\hat{\alpha}-\alpha = \frac{\sum_i S_i^2 \sum_i y_i - \sum_i S_i \sum_i S_i y_i}{n\sum_i S_i^2 - \left(\sum_i S_i\right)^2}. \tag{7.16}$$

In cases (7.11a) and (7.11b), according to (7.10) the terms of the numerator of (7.16) are both $O_p(n^{2d_x+d_y+5/2})$ and

$$n^{1/2-d_y}(\hat{\alpha}-\alpha) \xrightarrow{d} \frac{\int_0^1 X^2 d\xi\, Y(1) - \int_0^1 X d\xi \int_0^1 X dY}{\int_0^1 X^2 d\xi - \left(\int_0^1 X d\xi\right)^2}.$$

With $d_y < \frac{1}{2}$, the intercept is consistently estimated in these cases.

In cases (7.11c) and (7.11d), $d_y \le -d_x$ and the first term of the numerator of (7.16) is $O_p(n^{2d_x+d_y+5/2})$, while the second term is $O_p(n^{d_x+5/2}\log n)$ in case (7.11c) and $O_p(n^{d_x+5/2})$ in case (7.11d). In both cases, the first term is dominated by the second term. However, since the denominator of (7.16) is $O_p(n^{2d_x+3})$ the error of estimate is at worst $O_p(n^{-d_x-1/2}\log n)$ which is of small order with $d_x > -\frac{1}{2}$. The intercept is therefore consistently estimated in all cases.

7.3 Implications for Modelling

It is a commonplace in econometrics that in a stationary world an endogenous regressor results in inconsistent regression, whereas in a cointegrating world (due to the presence of unit roots) endogeneity may result in bias, but the regression is still consistent. The interesting question, answered precisely by considering fractionally integrated series, is what the transition between these two worlds looks like.

Referring to the discussion on page 10, consider an overdifferenced fractional process x with $d_x < -\frac{1}{2}$. If x is defined by (1.1)+(1.2) then

$$S_i = \sum_{j=1}^{i} x_j = \sum_{j=-\infty}^{i} b_j^* u_j \tag{7.17}$$

where $b_j^* = \sum_{k=\max\{0,1-j\}}^{i-j} b_k$, just as in (2.26). If $b_k \sim d_x k^{d_x-1}$ then as $i \to \infty$, $b_j^* \sim j^{d_S-1}$ as in (2.30), where $d_S = 1 + d_x < \frac{1}{2}$, so S_i is eventually a stationary long memory process and the limits in (7.10a) and (7.10b) do not apply. As explained in §7.1 these limits must be replaced by, respectively, $n^{-1}\sum_{i=1}^{n} S_i \to_{L_2} 0$ and $n^{-1}\sum_{i=1}^{n} S_i^2 \to_{L_2} \sigma_S^2$ where $\sigma_S^2 = \sigma_u^2 \sum_{j=0}^{\infty} b_j^{*2} < \infty$, as in (1.5).

Consider the case of model (7.8) with $\sigma_{uw} \neq 0$ and $d_x < -\frac{1}{2}$, so that necessarily $d_x + d_y < 0$. Instead of (7.15), the regression error of estimate now converges in mean square to the fixed limit σ_{Sy}/σ_S^2, where $\sigma_{Sy} = \sigma_{uw}\sum_{j=0}^{\infty} b_j^* c_j < \infty$ as in (4.3), matching (7.4) except with S replacing x. Thus, $\hat{\beta}$ is inconsistent and the solution for the asymptotic bias matches that for the model (7.1).

Thinking of the 'model space' as the set of the possible values of d_x that might generate the data, $-\frac{1}{2}$ is the point in model space representing the boundary between regions of stationarity and nonstationarity of the process S. The transition from a consistent albeit biased regression to inconsistency due to an endogenous regressor occurs, as may be expected, at the boundary of the stationarity region.

A striking feature of fractional theory is what might be called the relative fragility of antipersistence. Long memory processes are immune from the various difficulties due to singularities and deficiencies of shock process moments, as well as slower convergence rates, that have been detailed as attending the antipersistent case. These problems are evidently due to the relative proximity of the boundary in model space beyond which normalized partial sums are no longer asymptotically equicontinuous.

Consider the restriction $d_x + d_y > -\frac{1}{2}$, imposed by Assumption **6.9** as necessary for convergence to an a.s. continuous limit process, as shown in Theorem **6.13**. This is the constraint imposed when (unlike the previous example) the regressor in (7.8) is strictly exogenous, with $\sigma_{uw} = 0$ so that limit (7.11b) applies. A natural question to ask is why does this boundary matter in this case, and what actually happens if $d_x + d_y < -\frac{1}{2}$?

This question has a ready answer from Theorem **4.3** with S defined in (7.17) replacing x, noting that $d_x + d_y < -\frac{1}{2}$ is equivalent to $d_S + d_y < \frac{1}{2}$ since $d_S = 1 + d_x$. In (7.11b) the normalizing divisor has exponent smaller than $\frac{1}{2}$ and the sum consequently diverges, albeit having mean of zero. However, change the normalization of the numerator to $n^{-1/2}$ and according to Theorem **4.3**, the limit for (7.9) is

$$\sqrt{n}(\hat{\beta} - \beta) \xrightarrow{d} \mathrm{N}\left(0, \frac{V_{Sy}}{\sigma_S^4}\right) \tag{7.18}$$

having the same form as (7.3), except with S replacing x. But now consider the alternative scenario, in which the condition $d_S + d_y < \frac{1}{2}$ is violated in Theorem **4.3**. If $d_x + d_y \geq -\frac{1}{2}$, $V_{Sy} = \infty$ in equation (4.9). However, if $d_x + d_y > -\frac{1}{2}$ this is the same as $d_S + d_y > \frac{1}{2}$ and the data now satisfy the conditions of (7.11b). Suitably normalized, the sum of products has weak limit Ξ_{xy}.

In this instance, it is $d_x + d_y = -\frac{1}{2}$ that sets the boundary between the stationary regression world and the cointegrating world. The actual point of equality is a no-man's land where neither of the conditions hold, since both specify strict inequalities. These conditions, set respectively by Assumption **6.9** in Theorem **6.13** and Theorem **4.3**, turn out to be the two sides of the same coin.

Another question that is bound to be asked is whether there are circumstances in which the distribution of normalized $\hat{\beta} - \beta$ could be treated as mixed normal. In the case where $\sigma_{uw} = 0$, so that x_i is strongly exogenous under Assumption **4.1**, it is well known that in the unit root case this condition validates conventional large sample inference. Even when the error of estimate is not Gaussian in the limit the associated t statistic can be, although in this instance the normalizing divisor is not a constant but a random variable in the limit. In cases where strong exogeneity does not hold, techniques have been developed[1] for orthogonalizing the disturbances so that the regressor might be treated as conditionally fixed.

However, such methods appear incompatible with fractionally integrated disturbances. It is true that Ξ_1 as defined in (5.23) is mixed normal under strong exogeneity, and the same is true of Ξ_3 in (5.29). However, the conditioning variables are different and in the latter case are the lagged increments of the process Y itself. It is difficult to conceive of a transformation of $\Xi = \Xi_1 + \Xi_3$ having a tabulated distribution free of nuisance parameters.

[1] See [62], [69], and [75] inter alia.

7.4 Cointegration with Drift

The development in §7.2 specified that the regressor be free of drift, requiring it to be the partial sum of a zero-mean process. If this assumption is relaxed a different limit distribution is obtained, in which the trend dominates. In conventional unit root analysis the standard procedure is to partial out any drift by inclusion of a trend dummy in the regression, which results in yet another limit distribution but in this case one not dependent on the drift parameter.

The same approach can be followed in fractional cointegration. If the regressor increment process has the form $x_i + \mu$ where x_i is given as usual by (1.1)+(1.2), the partial sum has the form $S_i + \mu i$ for $i = 1, \ldots, n$. Removing dependence on μ is conveniently performed by demeaning and detrending the variable as a first stage and then running a second stage regression with the resulting residual as regressor. This is equivalent to the multiple regression including intercept and trend, according to the well-known Frisch-Waugh theorem. If S_i^* denotes the first-stage residual, the second stage error-of-estimate is just

$$\hat{\beta} - \beta = \frac{\sum_i S_i^* y_i}{\sum_i S_i^{*2}}. \tag{7.19}$$

The preliminary regression, of $S_i + \mu i$ onto intercept and trend, can be simplified by the approximation

$$\begin{bmatrix} n & \frac{1}{2}n(n+1) \\ \frac{1}{2}n(n+1) & \frac{1}{6}n(n+1)(2n+1) \end{bmatrix}^{-1} \sim \begin{bmatrix} 4n^{-1} & -6n^{-2} \\ -6n^{-2} & 12n^{-3} \end{bmatrix}.$$

An asymptotically equivalent form of the first-stage residual is then

$$S_i^* \sim S_i + \mu i - \left(\frac{4}{n} \sum_k (S_k + \mu k) - \frac{6}{n^2} \sum_k k(S_k + \mu k) \right) - \left(\frac{12}{n^3} \sum_k k(S_k + \mu k) - \frac{6}{n^2} \sum_k (S_k + \mu k) \right) i.$$

It can be verified that this series does not depend on μ in the limit as $n \to \infty$. Further elementary calculations show that

$$\sum_i S_i^{*2} \sim \sum_i S_i^2 - \frac{1}{n} \left(\sum_i S_i \right)^2 - \frac{12}{n^3} \left(\sum_i \left(i - \frac{n}{2} \right) S_i \right)^2$$

so that Theorem **3.2** and the continuous mapping theorem give

$$\frac{1}{n^{2+2d_x}}\sum_i S_i^{*2} \xrightarrow{d} \int_0^1 X^2 d\xi - \left(\int_0^1 X d\xi\right)^2 - 12\left(\int_0^1 (\xi - \tfrac{1}{2})X d\xi\right)^2. \quad (7.20)$$

Also, subject to the conditions of either (7.11a) or (7.11b),

$$\frac{1}{n^{1+d_x+d_y}}\sum_i S_i^* y_i \sim \frac{1}{n^{1+d_x+d_y}}\sum_i S_i y_i - \frac{1}{n^{3/2+d_x}}\sum_i S_i \frac{1}{n^{1/2+d_y}}\sum_i y_i$$
$$- \frac{12}{n^{5/2+d_x}}\sum_i \left(i - \frac{n}{2}\right) S_i \frac{1}{n^{3/2+d_y}}\sum_i \left(i - \frac{n}{2}\right) y_i$$
$$\xrightarrow{d} \int_0^1 X dY - \int_0^1 X d\xi Y(1) - 12\int_0^1 (\xi - \tfrac{1}{2}) X d\xi \left(\int_0^1 \xi dY - \tfrac{1}{2}Y(1)\right). \quad (7.21)$$

If either $d_x + d_y > 0$ or $\sigma_{uw} = 0$ and $d_x + d_y > -\frac{1}{2}$, $n^{1+d_x-d_y}(\hat{\beta} - \beta)$ in (7.19) converges in distribution to the ratio of the limits in (7.21) and (7.20).

Given the existence of the fBm X this development is identical to the usual unit root analysis, as treated in [13] among many other such references, in all respects except one. The exception is in the final term of (7.21), since the limit of the random sequence $n^{-3/2-d_y}\sum_i iy_i$ has not yet been examined. For the limit distribution in (7.21) to be well defined, this must be shown to have finite variance in the limit under the usual assumptions.

This development can conveniently follow the approach of §2.3 and §2.4. By analogy with (2.24), define an array $\{g_{ni}\}$ by

$$\sum_{i=1}^n iy_i = \sum_{i=1}^n i\sum_{l=0}^\infty c_l w_{i-l} = \sum_{i=-\infty}^n g_{ni}w_i \quad (7.22)$$

where if $c_l \sim d_y l^{d_y-1}$ it can be verified that

$$g_{ni} = \sum_{l=\max\{0,1-i\}}^{n-i} (l+i)c_l \sim d_y \sum_{l=\max\{0,1-i\}}^{n-i} (l^{d_y} + il^{d_y-1}).$$

To calculate the limiting mean square of (7.22), split the sum into the positive and nonpositive indices. For $0 < x \leq 1$,

$$g_{n[nx]} \sim d_y \int_0^{n-[nx]} \tau^{d_y} d\tau + [nx] d_y \int_0^{n-[nx]} \tau^{d_y-1} d\tau$$
$$\sim n^{d_y+1}\frac{d_y + x}{d_y + 1}(1-x)^{d_y} \quad (7.23)$$

whereas for $-\infty < x \le 0$,

$$
\begin{aligned}
g_{n[nx]} &\sim d_y \int_{-[nx]}^{n-[nx]} \tau^{d_y} \mathrm{d}\tau + [nx] d_y \int_{-[nx]}^{n-[nx]} \tau^{d_y-1} \mathrm{d}\tau \\
&\sim n^{d_y+1} \biggl(\frac{d_y}{d_y+1} \left((1-x)^{d_y+1} - (-x)^{d_y+1} \right) \\
&\qquad - (-x) \left((1-x)^{d_y} - (-x)^{d_y} \right) \biggr).
\end{aligned} \tag{7.24}
$$

Essentially, the requirement is to show that the functions of x in (7.23) and (7.24) are square-integrable. First, with $d_y > -\frac{1}{2}$,

$$
\int_0^1 \left(\frac{d_y + x}{d_y + 1} \right)^2 (1-x)^{2d_y} \mathrm{d}x = \frac{2d_y^2 + d_y + 1}{(d_y+1)(2d_y+1)(2d_y+3)} < \infty. \tag{7.25}
$$

Integrating the square of (7.24) over $(-\infty, 0]$ in closed form is not a simple exercise, but substituting the large-z approximation

$$
(1+z)^a - z^a = az^{a-1} + O(z^{a-2})
$$

for the terms in (7.24), first with $a = d_y + 1$ and then with $a = d_y$, shows that

$$
\frac{g_{n[nx]}}{n^{d_y+1}} = O((-x)^{d_y-1}) \tag{7.26}
$$

as $x \to -\infty$. When n is large enough, the normalized $\{g_{ni}\}$ sequence is square-summable. Similarly to Corollary **2.7**, applying Assumption **4.1** gives the result

$$
\lim_{n\to\infty} \mathrm{E} \left(\frac{1}{n^{3/2+d_y}} \sum_{i=1}^{n} i y_i \right)^2 = \lim_{n\to\infty} \frac{\sigma_w^2}{n^{3+2d_y}} \sum_{i=-\infty}^{n} g_{ni}^2 < \infty.
$$

This means that the limiting distribution in (7.21) is well defined. The contribution $\int_0^1 \xi \mathrm{d}Y(\xi)$ could doubtless be shown to be Gaussian by methods of the type developed in Chapters 2 and 3.

Chapter 8
Autocorrelated Shocks

The asymptotic analysis developed in Chapters 4 through 7 has been based on Assumption **4.1**, specifying serial independence of the shock variables u_i and w_i. Local dependence that can be removed by linear filtering is accounted for (see Theorem **1.4**), so that the assumption of zero autocorrelations for the shock process is not unreasonably restrictive. Nonetheless, Assumption **4.1** remains a special case and the introductory discussion in §3.5 (see page 51) suggests one reason at least why it might be useful to relax it. Except for the replacement of σ_u^2 by the long-run variance ω_u^2 in formulae, all but one of the limit results appearing in Chapters 4–6 under Assumption **4.1** continue to hold under the new assumptions to be set up in this chapter.

8.1 Correlation Analysis

Extending Assumption **1.2** into the multivariate context begins by defining some new symbols. Thus,

$$\gamma_{uw}(j) = \mathrm{E}(u_0 w_j) \text{ with } \omega_{uw} = \textstyle\sum_{j=-\infty}^{\infty} \gamma_{uw}(j)$$

and

$$\mu_{uw}^4(j,k,l) = \mathrm{E}(u_0 w_k u_j w_{j+l}) \text{ with } \varpi_{uw}^4 = \textstyle\sum_{j,k,l=-\infty}^{\infty} \mu_{uw}^4(j,k,l).$$

Under stationarity these moments are invariant to the time index and in particular, $\gamma_{uw}(0) = \sigma_{uw}$ and $\mu_{uw}^4(0,0,0) = \mu_{uw}^4$, where the latter symbols are defined in Assumption **4.1**(b).

8.1 Assumption

(a) $|d_x| < \frac{1}{2}$, $|d_y| < \frac{1}{2}$, and Assumption **1.2** holds for u_i and also for w_i with γ_w and ω_w^2 defined as for γ_u and ω_u^2 in Assumption **1.2**(a) and (1.4).

(b) $\gamma_{uw}(j) = O(|j|^{-1-\delta})$ for $\delta > 0$, and $|\omega_{uw}| < \infty$.

Asymptotics for Fractional Processes. James Davidson, Oxford University Press. © James Davidson (2025).
DOI: 10.1093/9780198955207.003.0008

(c) for $\delta > 0$,
  (i) $\mu^4_{uw}(j,k,l) = O(|k|^{-1-\delta}|l|^{-1-\delta})$
  (ii) $\mu^4_{uw}(j,k,l) - \gamma_{uw}(k)\gamma_{uw}(l) = O(|j|^{-1-\delta})$
  (iii) $\mu^4_{uw}(j,k,k+l) - \gamma_u(j)\gamma_w(j+l) = O(|k|^{-1-\delta})$
  and $|\varpi^4_{uw}| < \infty$. □

In the context of this assumption recall that $|0|^{-1-\delta}$ represents 1 in calculations where the counting index refers to a lag coefficient or order of autocovariance.

Case (i) of Assumption **8.1**(c) can be best appreciated by observing that under stationarity the expectations $\mathrm{E}(u_0 w_k u_j w_{j+l})$ are assumed to converge at the indicated rate to $\mathrm{E}(u_0 u_j w_{j+l})\mathrm{E}(w_0) = 0$ as $|k| \to \infty$ and to $\mathrm{E}(u_0 w_k u_j)\mathrm{E}(w_0) = 0$ as $|l| \to \infty$. According to (ii), the limit of $\mathrm{E}(u_0 w_k u_j w_{j+l})$ as $|j| \to \infty$ has the form $\mathrm{E}(u_0 w_k)\mathrm{E}(u_0 w_l)$ where the latter factors converge with k and l according to Assumption **8.1**(b). In case (iii), the limit of $\mathrm{E}(u_0 w_k u_j w_{j+k+l})$ as $|k| \to \infty$ is $\mathrm{E}(u_0 u_j)\mathrm{E}(w_0 w_{j+l})$. It is assumed implicitly in (i) that the divergences of the indices are not coordinated, by setting $k = l$ for example. This case is covered by (iii) with a given value of l. In the case $w_i = u_i$, the principal differences are that $\gamma_{uw} = \gamma_u = \gamma_w$ so that conditions (ii) and (iii) are equivalent and that, as in Assumption **4.1**, the shocks must possess a fourth moment.

The divergence rates in Assumption **8.1** are generally required to hold for any $\delta > 0$, in which case they are cited without comment. If δ must exceed some positive bound, to be specified, it is possible for δ to differ for u_i and w_i in Assumption **1.2**. In this case note that if either $\gamma_u(j) = 0$ or $\gamma_w(j) = 0$ for $j \neq 0$ then $\gamma_{uw}(j) = 0$ for $j \neq 0$ also, with a similar consideration for the fourth moments.

Summability of the various autocovariance sequences is a key property and the following pair of mechanical lemmas have a number of applications. The first one uses a technique that has already been applied in Theorem **1.4**. Keep in mind the useful identity

$$\sum_{k=1}^{j-1}(j-k)^{d-1}k^{-1-\delta} = \sum_{m=1}^{j-1} m^{d-1}(j-m)^{-1-\delta}$$

where $m = j - k$, hence, it doesn't matter which exponent plays which role in the sum.

8.2 Lemma If $d < 1$ and $\delta > 0$ then

$$\sum_{k=1}^{j-1}(j-k)^{d-1}k^{-1-\delta} \simeq j^{d-1}.$$

Proof Choose η from the interval $(1/(1+\delta), 1)$ and so write

$$\sum_{k=1}^{j-1}(j-k)^{d-1}k^{-1-\delta} = j^{d-1}\left(\frac{j-j^\eta}{j}\right)^{d-1}(A(j)+B(j))$$

where

$$A(j)+B(j) = \left(\sum_{k=1}^{[j^\eta]-1} + \sum_{k=[j^\eta]}^{j-1}\right)\left(\frac{j-k}{j-j^\eta}\right)^{d-1}k^{-1-\delta}$$

and $(j-j^\eta)/j \to 1$ as $j \to \infty$. Since $(j-k)/(j-j^\eta) > 1$ in $A(j)$ and $d-1<0$ and also $k^{-1-\delta}$ is summable, $A(\infty) < \infty$. Setting $m = j-k$ gives

$$\begin{aligned}
B(j) &= \sum_{m=1}^{j-[j^\eta]}\left(\frac{m}{j-j^\eta}\right)^{d-1}(j-m)^{-1-\delta} \\
&= (j-j^\eta)^{1-d}j^{-\eta(1+\delta)}\sum_{m=1}^{j-[j^\eta]} m^{d-1}\left(\frac{j-m}{j^\eta}\right)^{-1-\delta} \\
&\le (j-j^\eta)^{1-d}j^{-\eta(1+\delta)}\sum_{m=1}^{j-[j^\eta]} m^{d-1} \\
&\ll (j-j^\eta)j^{-\eta(1+\delta)} \le j^{1-\eta(1+\delta)} \to 0 \qquad (8.1)
\end{aligned}$$

as $j \to \infty$. The first inequality in (8.1) holds since $j - m \ge j^\eta$ so that in the sum over m the second factors are always smaller than 1. Replacing these by 1, the sum is then of $O((j-j^\eta)^d)$ by integral approximation. The final bound is of small order by choice of η. ∎

8.3 Lemma If $|d_x| < \frac{1}{2}$, $|d_y| < \frac{1}{2}$ and $\delta > 0$ then

$$\sum_{j=1}^{\infty}\sum_{k=1}^{\infty} j^{d_x-1}k^{d_y-1}|j-k|^{-1-\delta} < \infty. \qquad (8.2)$$

Proof Write the expression in (8.2) as

$$\sum_{j=1}^{\infty}\left(\sum_{k=1}^{j-1}+\sum_{k=j}^{\infty}\right)j^{d_x-1}k^{d_y-1}|j-k|^{-1-\delta} = A+B.$$

Both A and B are shown to be finite. First, by Lemma **8.2** with $d = d_y$,

$$A = \sum_{j=1}^{\infty} j^{d_x-1}\sum_{k=1}^{j-1}k^{d_y-1}(j-k)^{-1-\delta} \backsimeq \sum_{j=1}^{\infty} j^{d_x+d_y-2} < \infty.$$

Second, put $m = k - j$ to obtain

$$B = \sum_{j=1}^{\infty} j^{d_x-1} \sum_{k=j}^{\infty} k^{d_y-1}(k-j)^{-1-\delta} = \sum_{j=1}^{\infty} j^{d_x+d_y-2} \sum_{m=0}^{\infty} \left(\frac{j+m}{j}\right)^{d_y-1} m^{-1-\delta}$$
$$\ll \sum_{j=1}^{\infty} j^{d_x+d_y-2} < \infty$$

where the inequality holds since $(j+m)/j \geq 1$ and $d_y - 1 < 0$, while $m^{-1-\delta}$ is summable. ∎

In what follows, recall the definitions of x_i and y_i in (4.1). The first modification called for is to (4.3), which now becomes

$$\mathrm{E}(x_i y_i) = \sigma_{xy} = \sum_{j=0}^{\infty}\sum_{k=0}^{\infty} b_j c_k \gamma_{uw}(j-k). \tag{8.3}$$

This sum is finite if the autocovariances are summable.

8.4 Theorem Under Assumption **8.1**, $\sigma_{xy} < \infty$.

Proof Direct from Lemma **8.3**, given (8.3), specifications (4.1) and Assumption **8.1**(b). The summability criteria are invariant to the slowly varying components of b_j and c_k. ∎

These preliminaries lead to the next task, which is to prove the generalized version of Theorem **4.2**.

8.5 Theorem If Assumption **8.1** holds,

$$\frac{1}{n}\sum_{i=1}^{n} x_i y_i \xrightarrow{L_2} \sigma_{xy}.$$

Proof The expression that must be shown to vanish is (4.4). For brevity define $V_n = \mathrm{E}\left(n^{-1}\sum_{i=1}^{n} x_i y_i - \sigma_{xy}\right)^2$. Under stationarity of u_i and w_i,

$$\sum_{i=1}^{n}\sum_{k=1}^{n} \mathrm{E}(u_{i-j}w_{i-l}u_{k-m}w_{k-p}) = n\sum_{k=1}^{n} \mathrm{E}(u_0 w_{j-l} u_{k-m} w_{k-p})$$

and (4.4) can be written as

$$V_n = \frac{1}{n^2}\sum_{j=0}^{\infty}\sum_{m=0}^{\infty}\sum_{l=0}^{\infty}\sum_{p=0}^{\infty} b_j b_m c_l c_p$$
$$\times\, n\sum_{k=1}^{n}\left(\mu_{uw}^4(k-m, j-l, m-p) - \gamma_{uw}(j-l)\gamma_{uw}(m-p)\right).$$

Therefore, ignoring slowly varying components and applying Assumption **8.1**(c),

$$V_n \ll \frac{1}{n}\sum_{j=0}^{\infty}\sum_{m=0}^{\infty}\sum_{l=0}^{\infty}\sum_{p=0}^{\infty} j^{d_x-1}m^{d_x-1}l^{d_y-1}p^{d_y-1}$$

$$\times \sum_{k=1}^{n} |k-m|^{-1-\delta}|j-l|^{-1-\delta}|m-p|^{-1-\delta}. \tag{8.4}$$

Since $\sum_{j=0}^{\infty}\sum_{l=0}^{\infty} j^{d_x-1}l^{d_y-1}|j-l|^{-1-\delta} < \infty$ by Lemma **8.3**, these terms do not affect the order of magnitude and a statement equivalent to (8.4) is

$$V_n \ll \frac{1}{n}\sum_{k=1}^{n}\sum_{m=0}^{\infty} m^{d_x-1}|k-m|^{-1-\delta}\sum_{p=0}^{\infty} p^{d_y-1}|m-p|^{-1-\delta}. \tag{8.5}$$

Consider the sum over p in (8.5). Decompose this sum into the terms with $p < m$ and $p \geq m$ and in the second case set $q = p - m$ and rearrange, to get

$$\sum_{p=0}^{\infty} p^{d_y-1}|m-p|^{-1-\delta} = \sum_{p=0}^{m-1} p^{d_y-1}(m-p)^{-1-\delta} + m^{d_y-1}\sum_{q=0}^{\infty}\left(\frac{q+m}{m}\right)^{d_y-1} q^{-1-\delta}$$

$$\ll m^{d_y-1}. \tag{8.6}$$

This bound holds because the first term in (8.6) is bounded by Lemma **8.2** with $d = d_y$, while the sum in the second term is finite by summability of $q^{-1-\delta}$ since the other factor never exceeds 1. Substitute the bound in (8.6) back into (8.5) and again split the sum, this time into the terms with $m < k$ and with $m \geq k$. Apply Lemma **8.2** to the first of these components and since $k^{d_x+d_y-2}$ is summable, the result is

$$V_n \ll \frac{1}{n}\sum_{k=1}^{n}\left(\sum_{m=0}^{k-1} m^{d_x+d_y-2}(k-m)^{-1-\delta}\right.$$

$$\left. + k^{d_x+d_y-2}\sum_{m=k}^{\infty}\left(\frac{m}{k}\right)^{d_x+d_y-2}(m-k)^{-1-\delta}\right)$$

$$\ll \frac{1}{n}\sum_{k=1}^{n} k^{d_x+d_y-2} = O(n^{-1}). \quad \blacksquare$$

There is an instructive comparison here with the proof of Theorem **4.2**, which explicitly counted the nonzero terms of the sum V_n. The present argument works by showing that the nonnegligible terms of the sum are sparse enough under Assumption **8.1** to permit summability, without having to itemize them.

The next result to be generalized is Theorem **4.3**. The conditions here allow the shocks u and w to be weakly dependent processes but require mutual independence at all orders of lag.

8.6 Theorem If Assumption **8.1** holds, $d_x + d_y < \frac{1}{2}$, and shock processes $\{u_i\}$ and $\{w_i\}$ are independently distributed,

$$\frac{1}{\sqrt{n}}\sum_{i=1}^{n} x_i y_i \xrightarrow{d} \mathrm{N}(0, V_{xy})$$

where $V_{xy} < \infty$.

Proof As in Theorem **4.3**, the object is to show that $V_{xy} = E(\zeta^2) < \infty$ where $\zeta = \sum_{k=0}^{\infty}\sum_{j=0}^{\infty} b_k c_j Z(k,j)$ and $Z(k,j)$ is the weak limit of $n^{-1/2}\sum_{i=1}^{n} u_{i-k}w_{i-j}$, as in (4.8). Under the assumptions, the random variables $Z(k,j)$ have means of zero and can be shown to be Gaussian by a CLT for dependent data such as Theorem **3.10**.

Therefore, consider their covariances under the assumptions of the theorem. For any p such that $j + p + s$ and $k + p + t$ are both nonnegative, if the equalities $i - k = m - k - p$ and $i - j = m - j - p$ both hold then

$$\begin{aligned}\mathrm{E}(u_{i-k}w_{i-j}u_{m-k-p-t}w_{m-j-p-s}) &= \mathrm{E}(u_{i-k}u_{m-k-p-t})\mathrm{E}(w_{i-j}w_{m-j-p-s}) \\ &= \mathrm{E}(u_q u_{q-t})\mathrm{E}(w_r w_{r-s}) \\ &= \gamma_u(t)\gamma_w(s)\end{aligned}$$

where $q = i - k = m - p - k$ and $r = i - j = m - p - j$. By stationarity this expectation is the same for any i with $m = i + p$, and hence it is also the case that

$$\mathrm{E}\left(Z(k,j)Z(k+p+t, j+p+s)\right) = \gamma_u(t)\gamma_w(s).$$

Setting $b_k = 0$ for $k < 0$ and $c_j = 0$ for $j < 0$, define for each t and s

$$V_{xy}(t,s) = \gamma_u(t)\gamma_w(s)\sum_{p=-\infty}^{\infty}\left(\sum_{k=0}^{\infty} b_k b_{k+p+t}\sum_{j=0}^{\infty} c_j c_{j+p+s}\right).$$

The sums over k and j are respectively of $O(|p+t|^{2d_x-1})$ and $O(|p+s|^{2d_y-1})$ and for fixed t and s their product is summable over p according to the assumption $d_x + d_y < \frac{1}{2}$. Also, by Assumption **8.1**(a),

$$\sum_{t,s} |\gamma_u(t)\gamma_w(s)| = \sum_t |\gamma_u(t)| \sum_s |\gamma_w(s)| < \infty.$$

Hence, $V_{xy} = \mathrm{E}(\zeta^2) = \sum_{t,s} V_{xy}(t,s) < \infty$. ∎

8.2 The Covariance Decomposition

The decomposition of the partial sums in (4.13)–(4.15) depended critically on the ability to isolate the portions G_{1n} and G_{3n} having zero mean. Under Assumption **4.1** this worked exactly in finite samples. Autocorrelation of the shocks complicates matters and under Assumption **8.1** it is no longer the case that $\mathrm{E}(G_{1n}) = 0$ and $\mathrm{E}(G_{3n}) = 0$. The aim must be to see if a decomposition exists that provides a large-sample approximation to the same state of affairs.

To this end, let a sequence $\{L_n\}$ be chosen such that $L_n \to \infty$ but $L_n/n \to 0$ as $n \to \infty$. Then in place of (4.13)–(4.15) substitute the definitions

$$G_{1n} = \frac{1}{nK(n)} \sum_{i=1}^{n-1} \sum_{k=1}^{i} \sum_{j=0}^{\infty} b_j u_{k-j} \sum_{l=0}^{i+j-k-L_n} c_l w_{i+1-l} \tag{8.7}$$

$$G_{2n} = \frac{1}{nK(n)} \sum_{i=1}^{n-1} \sum_{k=1}^{i} \sum_{j=0}^{\infty} b_j u_{k-j} \sum_{v=-L_n}^{L_n} c_{i+j-k+1+v} w_{k-j+v} \tag{8.8}$$

$$G_{3n} = \frac{1}{nK(n)} \sum_{i=1}^{n-1} \sum_{k=1}^{i} \sum_{j=0}^{\infty} b_j u_{k-j} \sum_{l=i+j-k+2+L_n}^{\infty} c_l w_{i+1-l}. \tag{8.9}$$

Note that $G_n = G_{1n} + G_{2n} + G_{3n}$ as before, but L_n terms have been moved from each of G_{1n} and G_{3n} into G_{2n}. (An empty sum equals zero, note.) In view of Assumption **8.1**(b), under these definitions $\mathrm{E}(G_{1n}) = O(L_n^{-\delta})$ and $\mathrm{E}(G_{3n}) = O(L_n^{-\delta})$. With $L_n \sim n^{\alpha}$ for $\alpha < 1$, the convergence rate of $n^{-\alpha\delta}$ shows how the best choice of L_n might relate to the degree of dependence.

Expression (4.17) now has to be replaced by

$$\mathrm{E}(G_{2n}) = \frac{1}{nK(n)} \sum_{i=1}^{n-1} \sum_{k=0}^{\infty} a_{n,i-k}(i/n, 0) c_{k+1} F_n(k) + o(1) \tag{8.10}$$

where

$$F_n(k) = \sum_{v=-L_n}^{L_n} \frac{c_{k+1+v}}{c_{k+1}} \gamma_{uw}(v) \backsimeq \sum_{v=-L_n}^{L_n} \left(1 + \frac{v}{k}\right)^{d_y - 1} \gamma_{uw}(v).$$

For finite values of v, $(1 + v/k)^{d_y-1} = 1 + O(k^{-1})$. Assumption **8.1**(b) implies that only finite values of v contribute significantly to $F_n(k)$ and hence that $F_n(k) \to \omega_{uw}$ as $k \to \infty$ and $n \to \infty$. The counterpart expression to (4.18) (the terms of (8.10) with $k < i$) therefore takes the form

$$\frac{1}{nK(n)}\sum_{i=1}^{n-1}\sum_{k=0}^{i-1} a_{n,i-k}(i/n,0)c_{k+1}F_n(k) \sim \frac{d_y}{n^2}\sum_{i=1}^{n-1}\sum_{k=1}^{i}\left(\frac{k}{n}\right)^{d_x+d_y-1}F_n(k)$$

$$\to \omega_{uw}d_y\int_0^1\int_0^\tau \zeta^{d_x+d_y-1}d\zeta d\tau$$

$$= \frac{\omega_{uw}d_y}{(d_x+d_y)(1+d_x+d_y)} \tag{8.11}$$

as $n \to \infty$. The convergence of F_n can be asserted here, since the contribution of finite values of k to the limiting sum is of small order in n. This matches the limit in (4.18) except that ω_{uw} replaces σ_{uw}. Exactly the same modification applies to (4.19).

These considerations suffice to verify the following generalization of Theorem **4.4**.

8.7 Theorem Under Assumption **8.1** and $d_x + d_y > 0$, with G_{2n} defined by (8.8), $E(G_{2n}) \to \omega_{uw}\lambda_{xy}$ as $n \to \infty$ where λ_{xy} is defined in (4.16). □

The next result, explicitly invoking the autocorrelation structure of Assumption **8.1**(c), is the generalized form of Theorem **4.8**.

8.8 Theorem Under Assumption **8.1** and with G_{2n} defined by (8.8), $E(G_{2n} - E(G_{2n}))^2 = O(n^{-1})$.

Proof The argument in Theorem **4.8** is modified as follows. The term (4.46) is replaced by

$$P_{ik} = \sum_{j=0}^{\infty} b_j \sum_{v=-L_n}^{L_n} c_{i+1-k+j+v}(u_{k-j}w_{k-j+v} - \gamma_{uw}(v)).$$

The inequality (4.48) whose minorant appears in inequality (4.47) is therefore replaced by

$$|\mathrm{E}(P_{ik}P_{i-m,k-p})| \le \sum_{j=0}^{\infty}\left|\sum_{l=0}^{\infty} b_j b_l \sum_{v=-L_n}^{L_n}\sum_{v'=-L_n}^{L_n} c_{i+1-k+j+v}c_{i-m+1-k+p+l+v'}\right.$$
$$\left.\times\left(\mathrm{E}(u_{k-j}w_{k-j+v}u_{k-p-l}w_{k-p-l+v'}) - \gamma_{uw}(v)\gamma_{uw}(v')\right)\right|$$
$$\le \sum_{j=0}^{\infty}|b_j b_{j-p} c_{i+1-k+j}c_{i-m+1-k+j}H_n(j,i,k,m,p)| \tag{8.12}$$

where, setting $q = p + l - j$,

$$H_n(j,i,k,m,p) = \sum_{q=p-j}^{\infty}\frac{b_{j-p+q}}{b_{j-p}}\sum_{v=-L_n}^{L_n}\sum_{v'=-L_n}^{L_n}\frac{c_{i+1-k+j+v}}{c_{i+1-k+j}}\frac{c_{i-m+1-k+j+q+v'}}{c_{i-m+1-k+j}}$$
$$\times\left(\mu_{uw}^4(q,v,v') - \gamma_{uw}(v)\gamma_{uw}(v')\right). \tag{8.13}$$

The difference between (4.49) and (8.12) is therefore the replacement of $|\mu_{uw}^4 - \sigma_{uw}^2|$ by $|H_n(j,i,k,m,p)|$. The approach of Theorem **4.8** carries through unchanged provided this function is bounded in the limit as $n \to \infty$ for all values of its arguments, noting that these are liable to diverge with n.

According to Assumptions **8.1**(b) and (c),

$$\left|\sum_{q=-\infty}^{\infty}\sum_{v=-L_n}^{L_n}\sum_{v'=-L_n}^{L_n}\mu_{uw}^4(q,v,v') - \gamma_{uw}(v)\gamma_{uw}(v')\right| \to |\varpi_{uw}^4 - \omega_{uw}^2| < \infty$$

as $n \to \infty$. This summability implies that only finite values of v, v', and q appear in nonnegligible terms of the sum in (8.13). Attention therefore focuses on the various ratios of coefficients. For finite values of q, note that

$$\frac{b_{j-p+q}}{b_{j-p}} \backsimeq \left(1 + \frac{q}{j-p}\right)^{d_x-1} = 1 + O((j-p)^{-1}).$$

Similarly, for arbitrary argument r and finite values of v and q,

$$\frac{c_{r+q+v}}{c_r} \backsimeq \left(1 + \frac{v+q}{r}\right)^{d_y-1} = 1 + O(r^{-1}).$$

The cases in question are $r = i+1-k+j$ and $r = i-m+1-k+j$. One or other of these indices is diverging as $n \to \infty$ and either $r = O(n)$ or r relates to a collection of terms that are of small order in the normalized sum in (4.47). These facts establish that $H_n(j,i,k,m,p) = O(1)$ as each of its arguments diverges. If the majorant expression in (8.12) is decomposed into either B_{11} and B_{12}

as in (4.51) or into B_{21} and B_{22} as in (4.53), with $H_n(j,i,k,m,p)$ replacing $\mu^4_{uw} - \sigma^2_{uw}$ in each case, the bounds established in (4.56), (4.58), (4.60), and (4.62) all continue to apply. ∎

8.3 Stochastic Integrals

Having redefined G_{2n} in (8.8), it is next necessary to consider the effects of the corresponding changes in G_{1n} and G_{3n} following the analysis of Chapter 5. In the first case, considering the changes from (4.13) and (4.15) to (8.7) and (8.9), refer to the reorganization of the sum detailed in pages 89–90. The sequences of transformations described there, starting from (8.7) and (8.9), yield the results

$$G_{1n} = \frac{1}{\sqrt{n}} \sum_{m=-\infty}^{n-1} q_{n,m-L_n} w_{m+1}$$

and

$$G_{3n} = \frac{1}{\sqrt{n}} \sum_{p=-\infty}^{n-1} h_{n,p-L_n} u_p.$$

The time separations between u_p and w_m are now at least L_n periods, so that any correlation between the variables is rendered asymptotically negligible, according to Assumption **8.1**(b).

Changes need to be made to Theorems **6.3** and **6.4**. The modified version of Theorem **6.3**. is as follows.

8.9 Theorem Under Assumption **8.1**,

(i) $\lim_{n\to\infty} n^{-1}\mathrm{E}\left(\sum_{m=-nN}^{n-1}(q_{n,m-L_n} - q^N_{n,m-L_n})w_{m+1}\right)^2 = O(N^{2d_x-1})$

(ii) $\lim_{n\to\infty} n^{-1}\mathrm{E}\left(\sum_{m=-\infty}^{-nN-1} q_{n,m-L_n} w_{m+1}\right)^2 = O(N^{2(d_x+d_y-1)})$.

Proof This is by modifying the proof of Theorem **6.3**, specifically equation (6.11) for part (i) and equation (6.13) for part (ii). Consider the expression

$$F_n(m,p) = \sum_{r=-\infty}^{\infty} \frac{a_{n,m+r,p}}{a_{nmp}} \left(\sum_{q=-\infty}^{\infty} \frac{a_{n,m+r,p+q}}{a_{n,m+r,p}} \gamma_u(q) \right) \gamma_w(r) \qquad (8.14)$$

where γ_u and γ_w are the summable autocovariances of u and w, according to Assumption **8.1**(a). For convenience of notation, let it be assumed that $a_{n,m+r,p+q} = 0$ in (8.14) if $m+r$ and $p+q$ fall outside the respective ranges $[-Nn, n-1]$ and $(-\infty, -nN - L_n - 1]$, and also if $p+q = m+r$ or $m+r = n$.

The summability of the autocovariances means that only finite q and r play a nonnegligible role in the sum. If $m = [nt] = O(n)$ and $p = [ns] = O(n)$ and q and r are finite then with $d_y \geq 0$, it follows from Lemma **5.2** that $K_n^{-1} a_{n,m+r,p+q} \sim A(t,s)$. The equivalences (5.12) and (5.15) show that the individual terms in the sums are $O(n^{-1})$. The addition or deletion of a set of terms numbering $o(n)$ accordingly has a small-order effect on these normalized sums. It follows that each of the ratios in the terms of (8.14) is converging to 1 and hence that

$$F_n(m,p) \to \omega_u^2 \omega_w^2$$

as $n \to \infty$. In the case $d_y < 0$ the same conclusion follows from Lemma **5.4**, noting $K(n)^{-1} c^*_{[nv]} b_{[nu]} = O(n^{-1})$.

Assumption **8.1**(b) means that the shocks u_p and w_{m+1} can be treated as independent of one another in the limit. In view of (5.4) it is therefore possible to write

$$\begin{aligned}
\lim_{n\to\infty} \frac{1}{n} \mathrm{E}\left(\sum_{m=-nN}^{n-1} \left(q_{n,m-L_n} - q^N_{n,m-L_n}\right) w_{m+1} \right)^2 \\
&= \lim_{n\to\infty} \frac{1}{n^2 K(n)^2} \mathrm{E}\left(\sum_{m=-nN}^{n-1} \sum_{p=-\infty}^{-nN-L_n-1} a_{nmp} u_p w_{m+1} \right)^2 \\
&= \lim_{n\to\infty} \frac{1}{n^2 K(n)^2} \sum_{m=-nN}^{n-1} \sum_{p=-\infty}^{-nN-L_n-1} a^2_{nmp} F_n(m,p) \\
&\leq \omega_u^2 \omega_w^2 \int_{-N}^{1} \int_{-\infty}^{-N} \bar{A}^2(t,s) \mathrm{d}s \mathrm{d}t. \qquad (8.15)
\end{aligned}$$

If (8.15) replaces (6.11) in the proof of Theorem **6.3**(i), equation (6.12) completes the proof as before.

For part (ii), the proof of Theorem **6.3**(ii) is modified in just the same manner. In this case let $F_n(m,p)$ be defined so as to have terms vanish outside the range $(-\infty, -nN-1] \times (-\infty, m]$, for each m. With (6.13) modified similarly to (6.11), the same argument applies as before to obtain

$$\begin{aligned}
\lim_{n\to\infty} \frac{1}{n} \mathrm{E}\left(\sum_{m=-\infty}^{-nN-1} q_{n,m-L_n} w_{m+1} \right)^2 \\
&= \lim_{n\to\infty} \frac{1}{n^2 K(n)^2} \sum_{m=-\infty}^{-nN-1} \sum_{p=-\infty}^{m} a^2_{nmp} F_n(m,p) \\
&\leq \omega_u^2 \omega_w^2 \int_{-\infty}^{-N} \int_{-\infty}^{t} \bar{A}^2(t,s) \mathrm{d}s \mathrm{d}t.
\end{aligned}$$

The proof is completed by application of (6.14). ∎

The alternative version of Theorem **6.4** is proved in a precisely parallel manner, appealing to Lemmas **5.3** and **5.5**. The result will be stated for the record.

8.10 Theorem Under Assumption **8.1**,

$$\text{(i)}\ \lim_{n\to\infty} n^{-1}\mathrm{E}\left(\sum_{p=-nN}^{n-1}(h_{n,p-L_n} - h^N_{n,p-L_n})u_p\right)^2 = O(N^{2d_y-1})$$
$$\text{(ii)}\ \lim_{n\to\infty} n^{-1}\mathrm{E}\left(\sum_{p=-\infty}^{-nN-1} h_{n,p-L_n}u_p\right)^2 = O(N^{2(d_x+d_y-1)}). \qquad \square$$

8.4 Weak Convergence

The next step is to reconsider the weak convergence results of §6.3 and §6.4. Under Assumption **1.2**, the functional central limit theorem of Chapter 3 was proved with the approach of Theorem **3.18** replacing that of Theorem **3.2**. To reproduce the results of §6.3 under Assumption **8.1**, a parallel approach applies the method of Lemma **3.13** to establish the Gaussianity, in place of Lemma **3.4**.

Lemma 8.11 The conclusion of Lemma **6.12** holds under Assumption **8.1**.

Proof Similarly to Lemma **3.13**, Gaussianity is to be shown for the components $q_{1n}(t)$ and $q^N_{2n}(t)$ in (6.30) by testing the conditions of Theorem **3.10**. In this application, the role of the scale constants c_{ni} is played by the constants c_{np}, defined in (6.33) for the case of q_{1n} and in (6.40) for the case of q^N_{2n}. The main task is to verify the conditions of Assumption **3.9**. The same considerations arise as in the proof of Lemma **3.13** except that the dependence of the conditions on the sign of d in the univariate analysis now relate to d_x. Overlooking the nontrending factors attached to (6.33) and (6.40), the arguments in the two cases align closely.

Thus, if $d_x > 0$, since $\max_{1\le p\le[nt]} c_{np} = O(n^{-1/2})$ in (6.33), conditions (3.43) and (3.44) are satisfied for any $\alpha > 0$. If $d_x < 0$, $\max_{1\le p\le[nt]} c_{np} = O(n^{-1/2-d_x})$ in (6.33), but setting $\alpha > -2d_x$ in Assumption **3.9** meets the requirement. The form of (6.40) is an order-of-magnitude approximation to that of (3.53), but with M^k_{nj} defined similarly to (3.54) and c_{np} in (6.40) replacing c_{ni} in that formula,

$$M^k_{nj} \backsimeq \frac{((j-1)B_n + nk)^{d_x-1}}{n^{1/2+d_x}}. \tag{8.16}$$

In the cases $k > 0$,

$$\max_{1\le j\le r_n} M^k_{nj} = M^k_{n1} = O(n^{-3/2}) = o(B_n^{-1/2})$$

where the last equality holds for any $B_n = o(n)$. Also, in view of the square-summability of M^k_{nj} in (8.16) which holds for any $d_x < \frac{1}{2}$,

$$\sum_{j=1}^{r_n} (M^k_{nj})^2 = O(B_n^{-1}).$$

For the case $k = 0$, $\max_{1\le j\le r_n} M^0_{nj} = O(n^{-d_x-1/2})$ with $d_x < 0$, so here too setting $\alpha > -2d_x$ is required to satisfy (3.43). With these considerations, Gaussianity is established similarly to Lemma **3.13** for $q_{1n}(t)$ and the N components of $q^N_{2n}(t)$.

To extend this result to $q^N_n(t)$ requires a proof of independent increments, but this is supplied by Theorem **3.12** after making the substitution $i = p$ and hence $c_{ni} = c_{np}$, where c_{np} is defined variously as (6.33) and (6.40) or as (6.46) and (6.48). The remark following the theorem on page 55 applies analogously in the present case. ∎

8.12 Lemma The conclusion of Lemma **6.14** holds under Assumptions **8.1**, **6.9**(a), and **3.14**.

Proof To show this, two changes need to be made to the proof of Lemma **6.14**. The first is to invoke Corollary **3.15**, which must replace the mention of Theorem **3.5** in the cited proof of Theorem **6.13**. The second is to replace Theorem **A.8** by Theorem **3.16** (including the citation in Corollary **A.9**) to provide the proof of uniform integrability of maximal partial sums. ∎

8.13 Theorem The conclusions of Theorems **6.11** and **6.15** hold under Assumptions **8.1** and **6.9**.

Proof The proofs of Theorem **6.11** and the parallel Theorem **6.15** can be reiterated with the changes that Lemmas **8.11** and **8.12** are invoked in place of Lemmas **6.12** and **6.14**. ∎

The final consideration is of Theorems **6.16** and **6.17**, establishing convergence to stochastic integrals. Only one step in the proof of this result depended directly on the independence of the shocks. This was the L_2-approximation of G^N_{1n} by G^{N*}_{1n} in (6.68), for which the sequence k_n was required only to diverge more slowly than n. Now k_n might require further restriction, with Assumption **8.1** needing to be supplemented by a restriction on δ beyond positivity.

8.14 Theorem Under Assumptions **6.9** and **8.1**, $G^N_{1n} \to_d \Xi^N_1$ if in addition, setting $L_n \sim n^\alpha$ and $k_n \sim n^\mu$, there exist $\mu < 1$ and $\alpha < 1$ such that

$$\max\left\{\frac{1-\mu(d_x+\frac{1}{2})}{\alpha} - 1, \frac{1-2\mu(d_x+\frac{1}{2})}{1-\mu}\right\} < \delta \tag{8.17}$$

where δ is defined by Assumption **8.1**(b) and (c).

Proof Under functional form (8.7) the generic form of the covariance segments, replacing (6.63), is

$$G^N_{1n} = \frac{1}{nK(n)} \sum_{m=1-nN+L_n}^{n-1} \sum_{p=1-nN}^{m-L_n} a_{nmp} u_p w_{m+1}. \tag{8.18}$$

The decomposition in (6.65), after modification to ensure that the increments are independent in the limit, takes the form

$$G^{N*}_{1n} = \frac{1}{nK(n)} \sum_{j=1}^{k_n} \left(\sum_{p=n_0}^{n_{j-1}} a_{nn_{j-1}p} u_p \sum_{m=n_{j-1}+1+L_n}^{\min\{n,n_j+L_n\}} w_m \right).$$

The modified form of (6.67) is then

$$G^N_{1n} - G^{N*}_{1n} = \frac{1}{nK(n)} \sum_{j=1}^{k_n} \sum_{m=n_{j-1}+L_n+1}^{\min\{n,n_j+L_n\}-1} \left(\sum_{p=n_{j-1}+1}^{m-L_n} a_{nmp} u_p + \sum_{p=n_0}^{n_{j-1}} (a_{nmp} - a_{nn_{j-1}p}) u_p \right) w_{m+1}. \tag{8.19}$$

For convenience of notation let it be assumed that the length of sample available is actually $n + L_n$, recalling $n = n_{k_n}$ in the definitions following (6.63). Also note that m may be written m_j without ambiguity. In this setup, (6.68) may be replaced by $E(G^N_n - G^{N*}_n)^2 \le 2(A_n + B_n)$ where

$$A_n = \frac{1}{n^2K(n)^2} \sum_{j=1}^{k_n} \sum_{j'=1}^{k_n} \sum_{m_j=n_{j-1}+1+L_n}^{n_j-1+L_n} \sum_{p_j=n_{j-1}+1}^{m_j-L_n} a_{nm_jp_j} \times \sum_{m_{j'}=n_{j'-1}+1+L_n}^{n_{j'}-1+L_n} \sum_{p_{j'}=n_{j'-1}}^{m_{j'}-L_n} a_{nm_{j'}p_{j'}} E(u_{p_j} w_{m_j+1} u_{p_{j'}} w_{m_{j'}+1}) \tag{8.20}$$

and

$$B_n = \frac{1}{n^2 K(n)^2} \sum_{j=1}^{k_n} \sum_{j'=1}^{k_n} \sum_{m_j=n_{j-1}+1+L_n}^{n_j-1+L_n} \sum_{p_j=n_0}^{n_{j-1}} (a_{nm_jp_j} - a_{nn_{j-1}p_j})$$
$$\times \sum_{m_{j'}=n_{j'-1}+1+L_n}^{n_{j'}-1+L_n} \sum_{p_{j'}=n_0}^{n_{j'-1}} (a_{nm_{j'}p_{j'}} - a_{nn_{j'-1}p_{j'}}) \mathrm{E}(u_{p_j} w_{m_j+1} u_{p_{j'}} w_{m_{j'}+1}). \tag{8.21}$$

According to Assumption **8.1**(c),

$$\begin{aligned} \mathrm{E}(u_{p_j} w_{m_j+1} u_{p_{j'}} w_{m_{j'}+1}) &\leq \mu_{uw}^4(|p_j - p_{j'}|, m_j - p_j, m_{j'} - p_{j'}) \\ &= \gamma_{uw}(m_j - p_j)\gamma_{uw}(m_{j'} - p_{j'}) + ((\mu_{uw}^4(|p_j - p_{j'}|, m_j - p_j, m_{j'} - p_{j'}) \\ &\qquad - \gamma_{uw}(m_j - p_j)\gamma_{uw}(m_{j'} - p_{j'})) \\ &\ll L_n^{-2(1+\delta)} + |p_j - p_{j'}|^{-1-\delta}. \end{aligned} \tag{8.22}$$

The number of instances of expectation (8.22) appearing in (8.20) is of order $(n/k_n)^4$. Since $a_{nm_jp_j}/K(n) = O(1)$ and $|p_j - p_{j'}| \ll (n/k_n)|j - j'|$, it is possible to bound A_n as

$$A_n \ll \frac{k_n^2}{n^2}\left(\frac{n}{k_n}\right)^4 L_n^{-2(1+\delta)} + \frac{1}{n^2}\left(\frac{n}{k_n}\right)^{3-\delta} \sum_{j=1}^{k_n}\sum_{j'=1}^{k_n} |j - j'|^{-1-\delta}.$$

In the second of these terms, writing m for $|j - j'|$, in view of the summability of $m^{-1-\delta}$ it is found that

$$\sum_{j=1}^{k_n}\sum_{j'=1}^{k_n} |j - j'|^{-1-\delta} = \sum_{j=1}^{k_n}\left(\sum_{m=1}^{j-1} m^{-1-\delta} + \sum_{m=1}^{k_n-j} m^{-1-\delta}\right) \ll k_n. \tag{8.23}$$

Hence, after simplification, there is a bound on A_n of the form

$$A_n \ll \frac{n^2}{k_n^2} L_n^{-2(1+\delta)} + \frac{n^{1-\delta}}{k_n^{2-\delta}} = O(n^{\max\{C_1, C_2\}}) \tag{8.24}$$

where $C_1 = 2(1 - \mu - \alpha(1 + \delta))$ and $C_2 = 1 - \delta - \mu(2 - \delta)$.

In B_n, the double sum in (8.21) contains the product of two terms tagged j and j' respectively, for each pair (j, j'). Taking the jth term, apply the c_r inequality[1] with $r = 2$ and then, since $(n_{j-1} - n_0) = O(n)$ and $L_n = o(n)$,

[1] SLT Theorem 2.21.

Theorem **6.5**(ii) gives

$$\frac{1}{nK(n)}\sum_{m_j=n_{j-1}+1+L_n}^{n_j-1+L_n}\left|\sum_{p_j=n_0}^{n_{j-1}}(a_{nm_jp_j}-a_{nn_{j-1}p_j})\right|$$
$$\ll \sum_{m_j=n_{j-1}+1+L_n}^{n_j-1+L_n}\left(\frac{n_{j-1}-n_0}{n^2K(n)^2}\sum_{p_j=n_0}^{n_{j-1}}(a_{nm_jp_j}-a_{nn_{j-1}p_j})^2\right)^{1/2}$$
$$\ll \sum_{m_j=n_{j-1}+1+L_n}^{n_j-1+L_n}\left(\frac{m_j-n_{j-1}}{n}\right)^{d_x+1/2}$$
$$\ll \frac{(n_j-n_{j-1})^{d_x+3/2}}{n^{d_x+1/2}} \ll \frac{n}{k_n^{d_x+3/2}}.$$

The same bound holds for j', so for each pair (j,j') the products are bounded by a scale factor of order $n^2/k_n^{2d_x+3}$. Applying (8.23), in a similar way to (8.24) there is a bound on B_n of the form

$$B_n \ll k_n^2\frac{n^2}{k_n^{2d_x+3}}L_n^{-2(1+\delta)}+\frac{n^2}{k_n^{2d_x+3}}\left(\frac{n}{k_n}\right)^{-1-\delta}\sum_{j=1}^{k_n}\sum_{j'=1}^{k_n}|j-j'|^{-1-\delta}$$
$$= O(n^{\max\{D_1,D_2\}}) \tag{8.25}$$

where $D_1 = 2(1-\mu(d_x+\frac{1}{2})-\alpha(1+\delta))$ and $D_2 = 1-\delta-\mu(2d_x+1-\delta)$.

The bound in (8.25) dominates that in (8.24) for every $d_x < \frac{1}{2}$, so the proof is concluded by noting that (8.17) is the bound on δ as a function of μ and α that ensures both $D_1 < 0$ and $D_2 < 0$. ∎

In the usual way, the counterpart result for G_{3n}^N is stated for the record.

Theorem 8.15 $G_{3n}^N \to_d \Xi_3^N$ under Assumptions **6.9** and **8.1** if, similarly to Theorem **8.14**, there exist $\mu < 1$ and $\alpha < 1$ such that

$$\max\left\{\frac{1-\mu(d_y+\frac{1}{2})}{\alpha}-1,\ \frac{1-2\mu(d_y+\frac{1}{2})}{1-\mu}\right\} < \delta. \quad \square \tag{8.26}$$

As with Theorem **6.16**, the argument is identical following substitution of the complementary formulae, noting only that Theorem **6.6** has to be cited in place of Theorem **6.5**.

Theorems **8.14** and **8.15** are the only results in this chapter that may impose a positive lower bound on δ. However, conditions (8.17) and (8.26) should be viewed as no more than sufficient bounds on the amount of dependence permitted and are very likely to be stronger than necessary. First, note how the c_r inequality was needed to bound B_n in (8.21) to deal with the fact that the squared sums include $O(n^2)$ terms, when all but $O(n)$ of these terms are constrained to zero in Theorems **6.16** and **6.17**. In practice, most will be small. Another consideration is that the various values of δ specified by Assumption **8.1** need not be identical, whereas conditions (8.17) and (8.26) necessarily cite the largest of these, should they differ. In spite of these qualifications, if the processes are long memory with positive fractional parameter the conditions can be met for any $\delta > 0$, by setting α and μ close enough to 1. The necessity to constrain δ arises only in the antipersistent cases where d_x or d_y are negative.

Taken together, all the foregoing considerations make it possible to state the generalization of Theorem **6.10**, which is most simply given as follows.

8.16 Theorem Theorem **6.10** continues to hold without Assumption **4.1** if Assumption **8.1** holds for δ such that conditions (8.17) and (8.26) are satisfied. □

8.5 Variance Formulae

The final task is to generalize variance formulae (6.29) and (6.62), by application of the type of argument leading to (8.10) and (8.14).

8.17 Theorem If Assumption **8.1** holds, then (6.29) is replaced by

$$\mathrm{E}(Q^N(t)^2) = \begin{cases} \omega_u^2 \int_{-N}^t A^2(t,s)\mathrm{d}s, & d_y > 0 \\ \omega_u^2 \int_{-N}^t A^{*2}(t,s)\mathrm{d}s, & d_y < 0. \end{cases} \tag{8.27}$$

Proof The approach is similar to that of Theorem **8.9**. With $q_n^N(t)$ defined in (6.25),

$$\begin{aligned} \mathrm{E}(q_n^N(t)^2) &= \frac{1}{nK(n)^2}\mathrm{E}\left(\sum_{p=-nN}^{[nt]} a_{n[nt]p}u_p\right)^2 \\ &= \frac{1}{nK(n)^2}\sum_{p=-nN}^{[nt]} a_{n[nt]p}^2\left(\sigma_u^2 + 2\sum_{r=1}^{p+nN}\left(1 + \frac{a_{n,[nt],p-r} - a_{n[nt]p}}{a_{n[nt]p}}\right)\gamma_u(r)\right) \\ &= \frac{1}{nK(n)^2}\sum_{p=-nN}^{[nt]} a_{n[nt]p}^2\left(\omega_u^2 - 2\sum_{r=p+nN+1}^{\infty}\gamma_u(r)\right. \\ &\qquad\qquad \left. +2\sum_{r=1}^{p+nN}\frac{a_{n,[nt],p-r} - a_{n[nt]p}}{a_{n[nt]p}}\gamma_u(r)\right). \end{aligned} \tag{8.28}$$

Noting (5.12) and (5.15), Lemmas **5.2** and **5.4** imply that when r is finite,

$$\frac{a_{n,[nt],p-r} - a_{n[nt]p}}{a_{n[nt]p}} = O(1/n).$$

Therefore, since the $\gamma_u(r)$ are summable by Assumption **8.1**(a), the inner sums in the parentheses in the last member of (8.28) are both of small order as $n \to \infty$ and

$$\mathrm{E}(q_n^N(t)^2) = \frac{\omega_u^2}{nK(n)^2} \sum_{p=-nN}^{[nt]} a_{n[nt]p}^2 + o(1) \to \mathrm{E}(Q^N(t)^2)$$

as $n \to \infty$. The limits in (8.27) are found similarly to (6.29). ∎

The companion result for G_{3n}^N is stated as follows for completeness, the proof being left for the reader to supply.

8.18 Theorem If Assumption **8.1** holds, then (6.62) is replaced by

$$\mathrm{E}(H^N(s)^2) = \begin{cases} \omega_w^2 \int_{-N}^{s} E^2(s,t)\mathrm{d}t, & d_x > 0 \\ \omega_w^2 \int_{-N}^{s} E^{*2}(s,t)\mathrm{d}t, & d_x < 0. \end{cases}$$

□

Chapter 9
Frequency Domain Analysis

This chapter sets out the basics of an alternative approach to long memory analysis. The frequency domain is the favoured setting for nonparametric investigations of long memory since the parameter d can be estimated from the periodogram without any further assumptions about functional form. These techniques are well covered in the literature and will not be discussed here since the focus, as before, is on modelling long memory and the convergence of partial sums to fractional Brownian motion. Although the techniques of analysis are very different, the findings match those already obtained in the time domain, highlighting the fact that these are two complementary ways to study the same models. The modelling framework is somewhat limited in scope compared to the time domain, but some asymptotic results can be derived more easily and elegantly in the frequency context.

The chapter draws material from [18], joint work with Nigar Hashimzade and inspired by [11] among other sources. Other useful references include [10], [9], [82], [68], [26], [63], and [57]. A minor issue of notation arises in this chapter, because the frequent appearance of complex-valued terms threatens confusion with the use of the symbol i as the observation index in discrete time. The symbols t and s are therefore used in this context, with the symbol r representing a location in the time continuum.

9.1 Harmonizable Representation

The so-called harmonizable representation (equivalently, spectral representation) of a stationary stochastic process in the time domain assigns a distribution to random variations at different frequencies, instead of at different points in time. The source of the variations is taken to be an a.s. continuous, complex-valued process $U : [-\pi, \pi] \mapsto \mathbb{C}$, where $U(\lambda)$ for $\lambda \in [-\pi, \pi]$ denotes a process coordinate and $\overline{U}(\lambda)$ is its complex conjugate. The key feature of U is that it has orthogonal increments. Letting $\mathrm{d}\lambda$ denote an increment of the

Asymptotics for Fractional Processes. James Davidson, Oxford University Press. © James Davidson (2025).
DOI: 10.1093/9780198955207.003.0009

line and $\mathrm{d}U(\lambda)$ the variation of the process over this increment, the following properties are assumed, for $0 < \sigma_u^2 < \infty$:

$$\mathrm{d}\overline{U}(\lambda) = \mathrm{d}U(-\lambda) \tag{9.1a}$$

$$\mathrm{E}(\mathrm{d}U(\lambda)) = 0 \tag{9.1b}$$

$$\mathrm{E}(\mathrm{d}U(\lambda)\mathrm{d}\overline{U}(\mu)) = \begin{cases} \sigma_u^2 \mathrm{d}\lambda, & \lambda = \mu \\ 0, & \text{otherwise.} \end{cases} \tag{9.1c}$$

The process is symmetric about zero apart from the switch to the complex conjugate according to (9.1a). The two-sided domain $[-\pi, \pi]$ is in truth a mathematical convenience rather than a necessity, since one or the other half of it contains all the information about the process. Property (9.1c) implies, for the case $\lambda_2 > \lambda_1$,

$$\begin{aligned} \mathrm{E}|U(\lambda_2) - U(\lambda_1)|^2 &= \mathrm{E}\left(\int_{\lambda_1}^{\lambda_2} \mathrm{d}U(\lambda) \int_{\lambda_1}^{\lambda_2} \mathrm{d}\overline{U}(\mu)\right) \\ &= \sigma_u^2 \int_{\lambda_1}^{\lambda_2} \mathrm{d}\lambda = \sigma_u^2(\lambda_2 - \lambda_1). \end{aligned} \tag{9.2}$$

A case fulfilling the conditions of (9.1) has the form $U(\lambda) = \sigma_u(A(\lambda) + \mathrm{i}B(\lambda))/\sqrt{2}$ and $U(-\lambda) = \sigma_u(A(\lambda) - \mathrm{i}B(\lambda))/\sqrt{2}$ for $0 \leq \lambda \leq \pi$, where A and B are standard Brownian motions on the interval $[0, \pi]$ with $A(0) = B(0) = 0$. There is no explicit requirement in spectral theory that U be Gaussian, but a process with finite variance that is both a.s. continuous and has independent increments must also be Gaussian.[1]

The role played by U in defining the distribution of a time domain process is as the integrator function in a stochastic Stieltjes integral. The leading example is the harmonizable representation of a white noise time domain sequence $\{u_1, \ldots, u_n\}$, with mean zero and variance σ_u^2. This is connected to U via a Fourier transform, according to

$$u_t = \frac{1}{\sqrt{2\pi}} \int_{-\pi}^{\pi} \mathrm{e}^{\mathrm{i}t\lambda} \mathrm{d}U(\lambda), t = 1, \ldots, n. \tag{9.3}$$

At given time t, the oscillating function $\mathrm{e}^{\mathrm{i}t\lambda}$ selects for each λ variously large and small contributions from U, to be added to u_t. In this way the function

[1] See [6] Theorem 19.1, SLT Theorem 28.21.

in (9.3) contributes random variations to u_t at different frequencies. The identity

$$\int_{-\pi}^{\pi} \mathrm{e}^{\mathrm{i}t\lambda}\mathrm{d}U(\lambda) = \int_{-\pi}^{\pi} \mathrm{e}^{-\mathrm{i}t\lambda}\mathrm{d}\overline{U}(\lambda)$$

means that under the assumptions of (9.1), u_t in (9.3) and its moments are real-valued.

The domain of λ is bounded by $\pm\pi$ since π corresponds to the highest frequency over which variations can be observed in a sample of length n. The function $\mathrm{e}^{\mathrm{i}t\pi}$ oscillates $n/2$ times as t ranges from 1 to n, and to see higher frequencies requires a longer sample. Similarly the lowest observable frequency, with a half-cycle over the sample period, is $\lambda = \pi/n$.

The fact that $\{u_t\}$ in (9.3) forms an uncorrelated sequence is verified under (9.1) by

$$\begin{aligned} \mathrm{E}(u_t u_s) &= \frac{1}{2\pi}\mathrm{E}\left(\int_{-\pi}^{\pi} \mathrm{e}^{\mathrm{i}t\lambda}\mathrm{d}U(\lambda) \int_{-\pi}^{\pi} \mathrm{e}^{-\mathrm{i}s\mu}\mathrm{d}\overline{U}(\mu)\right) \\ &= \frac{\sigma_u^2}{2\pi}\int_{-\pi}^{\pi} \mathrm{e}^{\mathrm{i}(t-s)\lambda}\mathrm{d}\lambda = \begin{cases} \sigma_u^2, & t = s \\ 0, & \text{otherwise} \end{cases} \end{aligned} \tag{9.4}$$

(see (B.17)). The uncorrelatedness property maps directly via the Fourier transform from the uniform distribution of $\mathrm{d}U$ over frequencies, as demonstrated by (9.2), giving rise to the form of the expectation in (9.4). Also observe that the time series variance σ_u^2 here receives a new interpretation, matching the scale factor attached to U in (9.1c).

Given (9.3) as a starting point, an autocorrelated process is constructed by inserting a (in general) complex-valued function $h(\lambda)$, known as a transfer function or frequency response function, into the integral. The transfer function defines in effect a stochastic process $h(\lambda)U(\lambda)$ for $\lambda \in [-\pi, \pi]$, having heteroscedastic increments and so assigning greater or lesser variations, on average, to different frequencies. These map into different modes of autocorrelation under the Fourier transform.

To construct the harmonizable representation of a general moving average process $x_t = \varphi(B)u_t$ where $\varphi(B)$ is a lag polynomial of infinite order, define the transfer function

$$h(\lambda) = \varphi(\mathrm{e}^{-\mathrm{i}\lambda}) = \sum_{j=0}^{\infty} \varphi_j \mathrm{e}^{-\mathrm{i}j\lambda}. \tag{9.5}$$

Substitute from (9.3) so as to write, for $t = 1, \dots, n$,

$$x_t = \sum_{j=0}^{\infty} \varphi_j u_{t-j} = \frac{1}{\sqrt{2\pi}} \sum_{j=0}^{\infty} \varphi_j \int_{-\pi}^{\pi} \mathrm{e}^{\mathrm{i}(t-j)\lambda} \mathrm{d}U(\lambda)$$

$$= \frac{1}{\sqrt{2\pi}} \int_{-\pi}^{\pi} \mathrm{e}^{\mathrm{i}t\lambda} h(\lambda) \mathrm{d}U(\lambda). \tag{9.6}$$

Thus, in the linear framework h has the functional form of the lag polynomial with $\mathrm{e}^{-\mathrm{i}\lambda}$ taking the place of the lag operator B. As another example, the transfer function of a stationary ARMA(p, q) process $\phi(B)x_t = \theta(B)u_t$, written in solved form as $x_t = \varphi(B)u_t$ where $\varphi(B) = \theta(B)/\phi(B)$, is

$$h(\lambda) = \varphi(\mathrm{e}^{-\mathrm{i}\lambda}) = \frac{\theta(\mathrm{e}^{-\mathrm{i}\lambda})}{\phi(\mathrm{e}^{-\mathrm{i}\lambda})}. \tag{9.7}$$

More generally, any continuous function on $[0, \pi]$ can serve as a transfer function and a popular procedure (e.g., [27]) is to specify it semiparametrically.

To quantify the mapping on the real line, the squared modulus of h with suitable scaling factors defines the spectral density of the process,

$$f(\lambda) = \frac{\sigma_u^2}{2\pi} |h(\lambda)|^2. \tag{9.8}$$

Adapting the manipulation of (9.4), application of identity (B.2) gives the autocovariances as

$$\gamma_k = \mathrm{E}(x_t x_{t-k}) = \frac{\sigma_u^2}{2\pi} \int_{-\pi}^{\pi} \mathrm{e}^{\mathrm{i}k\lambda} |h(\lambda)|^2 \mathrm{d}\lambda$$

$$= 2 \int_0^{\pi} f(\lambda) \cos(k\lambda) \mathrm{d}\lambda, \; k = 0, 1, 2, \dots. \tag{9.9}$$

The better-known spectral density formula

$$f(\lambda) = \frac{\gamma_0}{2\pi} + \frac{1}{\pi} \sum_{j=1}^{\infty} \gamma_j \cos j\lambda \tag{9.10}$$

can be deduced by substituting it into (9.9) and using solution (B.18) to produce

$$2\int_0^\pi \left(\frac{\gamma_0}{2\pi} + \frac{1}{\pi}\sum_{j=1}^{\infty}\gamma_j \cos(j\lambda)\right)\cos(k\lambda)\mathrm{d}\lambda$$
$$= \frac{\gamma_0}{\pi}\int_0^\pi \cos(k\lambda)\mathrm{d}\lambda + \frac{2}{\pi}\sum_{j=1}^{\infty}\gamma_j \int_0^\pi \cos(j\lambda)\cos(k\lambda)\mathrm{d}\lambda$$
$$= \gamma_k.$$

For the linear moving average case in particular, (9.5) and (B.2) produce the expansion

$$|h(\lambda)|^2 = \varphi(\mathrm{e}^{-\mathrm{i}\lambda})\varphi(\mathrm{e}^{\mathrm{i}\lambda}) = \sum_{j=0}^{\infty}\varphi_j^2 + 2\sum_{k=1}^{\infty}\left(\sum_{j=0}^{\infty}\varphi_j\varphi_{j+k}\right)\cos(k\lambda). \quad (9.11)$$

Formulae (9.8) and (9.10) then give

$$\gamma_k = \sigma_u^2 \sum_{j=0}^{\infty}\varphi_j\varphi_{j+k}$$

providing an alternative derivation of (1.6).

9.2 The Fractional Model

The simplest case of long memory in this framework is the fractionally integrated moving average. If

$$x_t = \sum_{j=0}^{\infty} b_j u_{t-j} \quad (9.12)$$

where

$$b_j = \frac{\Gamma(d+j)}{\Gamma(d)\Gamma(j+1)} \quad (9.13)$$

as in (1.13), the transfer function has the form

$$h(\lambda) = \sum_{j=0}^{\infty} b_j \mathrm{e}^{-\mathrm{i}\lambda j} = (1-\mathrm{e}^{-\mathrm{i}\lambda})^{-d} \quad (9.14)$$

where the second equality is obtained by applying the generalized binomial expansion in (1.8). For this example the spectral density has the alternative forms (with the benefit of trigonometric identities (B.2) and (B.9))

$$f(\lambda) = \frac{\sigma_u^2}{2\pi}|1 - \mathrm{e}^{-\mathrm{i}\lambda}|^{-2d} = \frac{\sigma_u^2}{2^{d+1}\pi}(1 - \cos\lambda)^{-d} = \frac{\sigma_u^2}{2^{2d+1}\pi}(\sin\lambda/2)^{-2d}. \quad (9.15)$$

In view of the fact that $(2\sin\lambda/2)^2 = |\lambda|^2 + O(|\lambda|^4)$, $|h(\lambda)|^2$ is approximated by $|\lambda|^{-2d}$ at low frequencies with $|\lambda|$ close to zero. If $d > 0$, f diverges at the origin, a phenomenon characterizing all long memory processes having the properties described in Chapter 1. When $d < 0$ on the other hand, $f(0) = 0$, which is the spectral property characterizing antipersistence. By contrast, weakly dependent processes have spectral densities that are bounded at the origin and also, except in the antipersistent case, bounded away from zero at the origin.

The estimation of the spectral density is a central topic in time series analysis, but it is important not to lose sight of the fact that it measures only one aspect of a dynamic process. The model in (9.6) is said to be causal, because x_t reflects the arrow of time in being driven by present and past shocks while it is independent of future shocks. Contrast (9.14) with a noncausal, forward-looking moving average in which $b_j = \Gamma(d+j)/(\Gamma(d)\Gamma(1+j))$ is the coefficient of u_{t+j} for each $j > 0$, while the coefficient of u_{t-j} is zero. The transfer function for this model is $(1 - \mathrm{e}^{\mathrm{i}\lambda})^{-d}$ and the spectral density is also (9.15).

More dramatically, consider a symmetric two-sided moving average model (also noncausal)

$$x_t = \sum_{k=-\infty}^{\infty} b_k^* u_{t-k}$$

where $b_k^* = b_{-k}^* = \sum_{j=0}^{\infty} b_j b_{k+j}$. Calling the transfer function of this model $h^*(\lambda)$, it can be verified (compare (9.11)) that

$$h^*(\lambda) = \sum_{k=-\infty}^{\infty} b_k^* \mathrm{e}^{\mathrm{i}k\lambda} = b(\mathrm{e}^{-\mathrm{i}\lambda})b(\mathrm{e}^{\mathrm{i}\lambda}) = |h(\lambda)|^2.$$

In particular, in the fractionally integrated case with parameter $d/2$ so that $b_j = \Gamma(d/2 + j)/(\Gamma(d/2)\Gamma(j + 1))$, (9.14) gives

$$h^*(\lambda) = (1 - \mathrm{e}^{-\mathrm{i}\lambda})^{-d/2}(1 - \mathrm{e}^{\mathrm{i}\lambda})^{-d/2} = |1 - \mathrm{e}^{-\mathrm{i}\lambda}|^{-d}.$$

The spectral density of this model is also identical with (9.15). In this case the transfer function is real, not complex, which is always the case with time-symmetric models, but this information is lost in taking the modulus.

The implication of these examples is that spectral densities contain no information about time ordering and directions of causation. However, one very useful feature of $f(\lambda)$ is the connection via (9.9) to the autocovariance sequence, which is likewise invariant to the direction of causality. Here, for the case of (9.14), is the harmonic counterpart of the time domain derivation of γ_k in Theorem **1.3**.

9.1 Theorem If the transfer function is (9.14) with $|d| < \frac{1}{2}$,

$$\gamma_k = \sigma_u^2 \frac{\Gamma(1-2d)\Gamma(d+k)}{\Gamma(1-d+k)} \frac{\sin \pi d}{\pi}. \tag{9.16}$$

Proof Inserting (9.15) into formula (9.9) gives

$$\gamma_k = \frac{\sigma_u^2}{2^{2d}\pi} \int_0^\pi (\sin \lambda/2)^{-2d} \cos(k\lambda) d\lambda. \tag{9.17}$$

The integral solution in (B.19) with $x = \lambda/2$, $\nu = 1 - 2d$, and $a = 2k$, using (B.14) and also noting $\cos(k\pi) = (-1)^k$ for integer k by (B.4), produces

$$\begin{aligned} \int_0^\pi (\sin \lambda/2)^{-2d} \cos(k\lambda) d\lambda &= \frac{\pi\Gamma(2-2d)(-1)^k}{2^{-2d}(1-2d)\Gamma(1-d+k)\Gamma(1-d-k)} \\ &= \frac{\Gamma(1-2d)\Gamma(d+k)}{2^{-2d}\Gamma(1-d+k)} \sin \pi d \end{aligned} \tag{9.18}$$

where the second equality uses (B.13), (B.15), and then (B.5). The proof is completed by substituting (9.18) into (9.17). ∎

9.3 The Partial Sum Process

The normalized partial sum of the fractional process (9.12) has harmonizable representation as follows, after substituting from (9.6) and (9.14) and resolving the sum of the terms $e^{it\lambda}$ as a geometric series. For $r \in [0, 1]$,

$$
\begin{aligned}
X_n(r) &= \frac{1}{n^{d+1/2}} \sum_{t=1}^{[nr]} x_t \\
&= \frac{1}{n^{d+1}\sqrt{2\pi/n}} \int_{-\pi}^{\pi} \sum_{t=1}^{[nr]} \mathrm{e}^{\mathrm{i}t\lambda}(1-\mathrm{e}^{-\mathrm{i}\lambda})^{-d} \mathrm{d}U(\lambda). \\
&= \frac{1}{n^{d+1}\sqrt{2\pi/n}} \int_{-\pi}^{\pi} \frac{\mathrm{e}^{\mathrm{i}([nr]+1)\lambda} - \mathrm{e}^{\mathrm{i}\lambda}}{\mathrm{e}^{\mathrm{i}\lambda}-1}(1-\mathrm{e}^{-\mathrm{i}\lambda})^{-d} \mathrm{d}U(\lambda). \qquad (9.19)
\end{aligned}
$$

The question of interest is, how should this formulation behave as the sample size increases, to complement the time domain device of mapping integer dates into a continuum? The trick is to make a change of variable in the integral from λ to λ/n. The function $\mathrm{e}^{\mathrm{i}t\pi/n}$ oscillates just half a cycle as t ranges from 1 to n, and changing the range of the integral from $[-\pi, \pi]$ to $[-n\pi, n\pi]$ extends the domain of U to accommodate the higher frequencies observable as n increases. Letting $X(r)$ denote the limiting case of $X_n(r)$, this is found heuristically as follows, making the change of variable in (9.19), rearranging, and letting $n \to \infty$.

$$
X_n(r) = \frac{1}{\sqrt{2\pi}} \int_{-n\pi}^{n\pi} \frac{\mathrm{e}^{\mathrm{i}([nr]+1)\lambda/n} - \mathrm{e}^{\mathrm{i}\lambda/n}}{n(\mathrm{e}^{\mathrm{i}\lambda/n}-1)} \frac{(1-\mathrm{e}^{-\mathrm{i}\lambda/n})^{-d}}{n^d} \mathrm{d}U(\lambda) \to X(r)
$$

where

$$
X(r) = \frac{1}{\sqrt{2\pi}} \int_{-\infty}^{\infty} \frac{\mathrm{e}^{\mathrm{i}\lambda r}-1}{\mathrm{i}\lambda}(\mathrm{i}\lambda)^{-d} \mathrm{d}U(\lambda). \qquad (9.20)
$$

This argument does not amount to a proof of weak convergence, but the limit formula shows the harmonizable representation of fractional Brownian motion. If U is Gaussian, such proofs do not involve a central limit theorem as such and are a matter of showing uniform tightness of the sequence of measures and L_2-convergence of the increments. An example is Theorem 2.2 of [18].

The spectral density of this continuous-time process is $\sigma_u^2|\lambda|^{-2d}/2\pi$. Just as the generalized binomial form (1.13) is not the only model that converges to (2.1), so the limit in (9.20) features a transfer function that is common to all long memory models in the neighbourhood of zero, not only (9.14). However, to show convincingly that (9.20) is indeed the harmonizable representation of (2.1), it needs at least to be shown that apart from optional scale factors, the increment variances match as functions of d. This is done as follows.

9.2 Theorem If X is the limit process having the harmonizable representation in (9.20), with $|d| < \frac{1}{2}$ and (9.1) holding, then

$$\mathrm{E}(X(r+\delta) - X(r))^2 = \frac{\sigma_u^2 \Upsilon_d}{\Gamma(d+1)^2}\delta^{2d+1} \tag{9.21}$$

where Υ_d is defined in (2.6).

Proof Using (9.20), (9.4), (B.2), and (B.9),

$$\begin{aligned}
\mathrm{E}(X(r+\delta) - X(r))^2 &= \frac{1}{2\pi}\mathrm{E}\left(\int_{-\infty}^{\infty} \mathrm{e}^{\mathrm{i}\lambda r}\frac{\mathrm{e}^{\mathrm{i}\lambda\delta}-1}{\mathrm{i}\lambda}(\mathrm{i}\lambda)^{-d}\mathrm{d}U(\lambda)\right. \\
&\qquad \left.\times \int_{-\infty}^{\infty} \mathrm{e}^{-\mathrm{i}\mu r}\frac{\mathrm{e}^{-\mathrm{i}\mu\delta}-1}{-\mathrm{i}\mu}(-\mathrm{i}\mu)^{-d}\mathrm{d}\overline{U}(\mu)\right) \\
&= \frac{\sigma_u^2}{\pi}\int_0^{\infty} |\mathrm{e}^{\mathrm{i}\lambda\delta} - 1|^2|\lambda|^{-2d-2}\mathrm{d}\lambda \\
&= \frac{4\sigma_u^2}{\pi}\int_0^{\infty} \sin^2(\lambda\delta/2)\lambda^{-2d-2}\mathrm{d}\lambda.
\end{aligned} \tag{9.22}$$

Setting $\mu = -2d - 1$ and $a = \delta/2$ in (B.21), also noting that $\cos(-\pi d - \pi/2) = -\sin(\pi d)$ by (B.6), yields the result

$$\begin{aligned}
\int_0^{\infty} \sin^2(\lambda\delta/2)\lambda^{-2d-2}\mathrm{d}\lambda &= \frac{\Gamma(-2d-1)\sin(\pi d)\delta^{2d+1}}{2} \\
&= \frac{\pi\delta^{2d+1}}{4\Gamma(2d+2)\cos(\pi d)}
\end{aligned} \tag{9.23}$$

where the second equality applies successively (B.15) with $x = -2d - 1$, then (B.8), (B.6), and (B.7). Substituting into (9.22) gives

$$\mathrm{E}(X(r+\delta) - X(r))^2 = \frac{\sigma_u^2\delta^{2d+1}}{\Gamma(2d+2)\cos(\pi d)}. \quad \blacksquare \tag{9.24}$$

Applying (2.13) to (9.24) shows that this expression matches (9.21) when Υ_d is given by (2.6). Alternatively, a direct match is found using formula (2.15). Remarkably, the time domain formula in (2.4) appears at no stage in this derivation.

Recalling the remarks on page 15, note the reason for the difference between Theorem **9.2** and Theorem **2.1**, comparing in particular (9.24) with (2.15). The transfer function (9.14), appearing in (9.19) and incorporated into the limit process (9.20), is derived from the assumption that the moving average coefficients have the form (9.13). The interesting fact, which is apparent from Theorem **2.6** and (2.26), is that to change Υ_d in (2.15) to $\Upsilon_d/(d+1)^2$ so that Theorems **2.1** and **9.2** give matching formulae is a matter of giving b_j the form $dj^{d-1}L(j)/\Gamma(d+1)$ instead of (1.2). Although (9.20) is built on a specific choice of transfer function, it is only in the matter of this scale factor that the time domain and frequency domain formalizations can be distinguished in the limit.

A more significant difference between the representations is to be found in the comparison of Corollary **2.8** and Theorem **2.10**. The harmonic framework cannot accommodate a nonparametric dependence setup of the type captured by Assumption **1.2**(b). Short-run dynamics can enter only via the transfer function. For example, the process defined in (1.23) has $h(\lambda) = (1 - \mathrm{e}^{-\mathrm{i}\lambda})^{-d}\varphi(\mathrm{e}^{-\mathrm{i}\lambda})$ where $\varphi(\mathrm{e}^{-\mathrm{i}\lambda})$ is defined in (9.7). Paralleling the development in (9.19) and (9.20), since $\varphi(\mathrm{e}^{-\mathrm{i}\lambda/n}) \to \varphi(1)$ as $n \to \infty$ when the lag coefficients are summable, when $x_t = (1 - B)^{-d}\varphi(B)u_t$ with u_t from (9.3) the limiting case of the partial sum process as $n \to \infty$ is found as

$$X_n(r) = \frac{1}{n^{d+1/2}}\sum_{t=1}^{[nr]} x_t \to X(r) = \frac{\varphi(1)}{\sqrt{2\pi}}\int_{-\infty}^{\infty} \frac{\mathrm{e}^{\mathrm{i}\lambda r} - 1}{\mathrm{i}\lambda}(\mathrm{i}\lambda)^{-d}\mathrm{d}U(\lambda).$$

Similarly to what was shown in Theorem **1.4**, the only effect of the weakly dependent shocks on the limit distribution is the possible change of scale.

9.4 Covariance Analysis

The next step is to study the relationships between different fractional processes. Let U and W be a pair of frequency domain processes on $(-\infty, +\infty)$ where U satisfies (9.1) and W the counterpart condition with variance $0 < \sigma_w^2 < \infty$. Further suppose that X and Y are fractional Brownian motions with harmonic representations

$$X(r) = \frac{1}{\sqrt{2\pi}}\int_{-\infty}^{\infty} \frac{\mathrm{e}^{\mathrm{i}\lambda r} - 1}{\mathrm{i}\lambda}(\mathrm{i}\lambda)^{-d_x}\mathrm{d}U(\lambda) \qquad (9.25)$$

and

$$Y(r) = \frac{1}{\sqrt{2\pi}} \int_{-\infty}^{\infty} \frac{e^{i\mu r} - 1}{i\mu} (i\mu)^{-d_y} dW(\mu) \tag{9.26}$$

for $r \in [0, 1]$, as in (9.20). The additional condition required is

$$E(dU(\lambda)d\overline{W}(\mu)) = \begin{cases} \sigma_{uw} d\lambda, & \lambda = \mu \\ 0, & \text{otherwise} \end{cases} \tag{9.27}$$

where σ_{uw} is the covariance linking these variables.

The covariance of contemporaneous increments is found by a variation of Theorem **9.2** with an additional trick based on two useful identities. Since $e^{i\pi/2} = i$ and $e^{-i\pi/2} = -i$, it follows that

$$(i\lambda)^{-d} = |\lambda|^{-d} e^{-i\pi d \mathrm{sgn}(\lambda)/2} \tag{9.28}$$

and

$$(-i\lambda)^{-d} = |\lambda|^{-d} e^{i\pi d \mathrm{sgn}(\lambda)/2} \tag{9.29}$$

where $\mathrm{sgn}(\lambda)$ denotes the sign of λ, either $+1$ or -1. This function has the convenient property of being constant over the positive and negative parts of the domain and hence can be factored out of each side of an integral, which is especially convenient if the remaining part of the integrand is real valued.

9.3 Theorem If $|d_x| < \frac{1}{2}$, $|d_y| < \frac{1}{2}$, and (9.27) holds,

$$E(X(r + \delta) - X(r))Y((r + \delta) - Y(r))) = \frac{\sigma_{uw} \cos(\pi(d_x - d_y)/2)\delta^{d_x+d_y+1}}{\Gamma(d_x + d_y + 2)\cos(\pi(d_x + d_y)/2)}. \tag{9.30}$$

Proof Following a manipulation similar to (9.4), applying (B.2) twice and then (B.9) produces

$$
\begin{aligned}
&\mathrm{E}(X(r+\delta)-X(r))Y((r+\delta)-Y(r))) \\
&= \frac{1}{2\pi}\mathrm{E}\left(\int_{-\infty}^{\infty} \mathrm{e}^{\mathrm{i}\lambda r}\frac{\mathrm{e}^{\mathrm{i}\lambda\delta}-1}{\mathrm{i}\lambda}(\mathrm{i}\lambda)^{-d_x}\mathrm{d}U(\lambda)\right. \\
&\qquad\qquad \left.\times\int_{-\infty}^{\infty} \mathrm{e}^{-\mathrm{i}\mu r}\frac{\mathrm{e}^{-\mathrm{i}\mu\delta}-1}{-\mathrm{i}\mu}(-\mathrm{i}\mu)^{-d_y}\mathrm{d}\overline{W}(\mu)\right) \\
&= \frac{\sigma_{uw}}{2\pi}\int_{-\infty}^{\infty} |\mathrm{e}^{\mathrm{i}\lambda\delta}-1|^2|\lambda|^{-d_x-d_y-2}\mathrm{e}^{-\mathrm{i}\pi\mathrm{sgn}(\lambda)(d_x-d_y)/2}\mathrm{d}\lambda \\
&= \frac{\sigma_{uw}}{2\pi}(\mathrm{e}^{-\mathrm{i}\pi(d_x-d_y)/2}+\mathrm{e}^{\mathrm{i}\pi(d_x-d_y)/2})\int_0^{\infty} |\mathrm{e}^{\mathrm{i}\lambda\delta}-1|^2\lambda^{-d_x-d_y-2}\mathrm{d}\lambda \\
&= \frac{4\sigma_{uw}}{\pi}\cos(\pi(d_x-d_y)/2)\int_0^{\infty}\sin^2(\lambda\delta/2)\lambda^{-d_x-d_y-2}\mathrm{d}\lambda. \qquad (9.31)
\end{aligned}
$$

The next step is to solve the integral in (9.31) with the help of (B.21). Then, successively applying (B.15), (B.8), and (B.7) finally yields (9.30). ∎

In the case $d_x = d_y$, the formula found for the univariate case in (9.24) is identical to (9.30) except for the replacement of σ_{uw} by σ_u^2. It will also not escape notice that (9.30) can also be written as

$$
\frac{\sigma_{uw}\Upsilon_{xy}\delta^{d_x+d_y+1}}{\Gamma(d_x+1)\Gamma(d_y+1)}
$$

where Υ_{xy} is defined in (4.36) and also in (2.54), and so matches the formula in (2.53) apart from the choice of specification discussed in the previous section.

9.5 Stochastic Integral

The next result, for comparison with the time domain calculation leading to (4.23), is the expected value of the stochastic integral of X with respect to Y. The differential increment of Y is found from the harmonic representation (9.26) as

$$
\begin{aligned}
\mathrm{d}Y(r) &= \frac{1}{\sqrt{2\pi}}\int_{-\infty}^{\infty}\frac{\partial}{\partial r}\frac{\mathrm{e}^{\mathrm{i}\mu r}-1}{\mathrm{i}\mu}(\mathrm{i}\mu)^{-d_y}\mathrm{d}W(\mu)\mathrm{d}r \\
&= \frac{1}{\sqrt{2\pi}}\int_{-\infty}^{\infty} e^{\mathrm{i}\mu r}(\mathrm{i}\mu)^{-d_y}\mathrm{d}W(\mu)\mathrm{d}r. \qquad (9.32)
\end{aligned}
$$

Teaming this formula with (9.25) and forming the integral with respect to time gives

$$\int_0^1 X\mathrm{d}Y = \frac{1}{2\pi}\int_0^1 \left(\int_{-\infty}^{\infty} \frac{\mathrm{e}^{\mathrm{i}\lambda r}-1}{\mathrm{i}\lambda}(\mathrm{i}\lambda)^{-d_x}\mathrm{d}U(\lambda)\right.$$
$$\left.\times \int_{-\infty}^{\infty} \mathrm{e}^{-\mathrm{i}\mu r}(-\mathrm{i}\mu)^{-d_y}\mathrm{d}\overline{W}(\mu)\right)\mathrm{d}r. \quad (9.33)$$

9.4 Theorem If (9.27) holds, $|d_x| < \frac{1}{2}$, $|d_y| < \frac{1}{2}$, and $d_x + d_y > 0$,

$$\mathrm{E}\left(\int_0^1 X\mathrm{d}Y\right) = \sigma_{uw}\frac{\Gamma(1-d_x-d_y)\sin\pi d_y}{\pi(1+d_x+d_y)(d_x+d_y)}. \quad (9.34)$$

Proof Substituting from (9.28) and (9.29) and applying (9.27), the expected value of (9.33) has the form

$$\mathrm{E}\left(\int_0^1 X\mathrm{d}Y\right) = \frac{\sigma_{uw}}{2\pi}\int_0^1\int_{-\infty}^{\infty}\frac{1-\mathrm{e}^{-\mathrm{i}\lambda r}}{\mathrm{i}\lambda}|\lambda|^{-d_x-d_y}(\mathrm{e}^{\mathrm{i}\pi\mathrm{sgn}(\lambda)(d_y-d_x)/2})\mathrm{d}\lambda\mathrm{d}r. \quad (9.35)$$

To evaluate the double integral in (9.35), first make a change of variable $\nu = \lambda r$. Since $0 \le r \le 1$, $\mathrm{sgn}(\lambda) = \mathrm{sgn}(\nu)$ and (9.35) factorizes into two integrals, where the second one can be split into positive and negative regions over which $\mathrm{sgn}(\nu)$ is constant. Thus,

$$\begin{aligned}\mathrm{E}\left(\int_0^1 X\mathrm{d}Y\right) &= \frac{\sigma_{uw}}{2\pi}\int_0^1 r^{d_x+d_y}\mathrm{d}r\int_{-\infty}^{\infty}\frac{1-\mathrm{e}^{-\mathrm{i}\nu}}{\mathrm{i}\nu}|\nu|^{-d_x-d_y}\mathrm{e}^{-\mathrm{i}\pi\mathrm{sgn}(\nu)(d_x-d_y)/2}\mathrm{d}\nu \\ &= \frac{\sigma_{uw}}{2\pi(1+d_x+d_y)}\left(\mathrm{e}^{-\mathrm{i}\pi(d_x-d_y)/2}J + \mathrm{e}^{\mathrm{i}\pi(d_x-d_y)/2}\overline{J}\right) \end{aligned} \quad (9.36)$$

where, since $1-\mathrm{e}^{-\mathrm{i}\nu} = 1-\cos\nu + \mathrm{i}\sin\nu$ from (B.1),

$$\begin{aligned} J &= \int_0^{\infty}\frac{1-\mathrm{e}^{-\mathrm{i}\nu}}{\mathrm{i}\nu}\nu^{-d_x-d_y}\mathrm{d}\nu \\ &= -\mathrm{i}\int_0^{\infty}\frac{1-\cos\nu}{\nu^{1+d_x+d_y}}\mathrm{d}\nu + \int_0^{\infty}\frac{\sin\nu}{\nu^{1+d_x+d_y}}\mathrm{d}\nu. \end{aligned}$$

To resolve the first term of J, first apply (B.9) and then (B.21) with $\mu = -d_x-d_y$ and $a = \frac{1}{2}$, followed by (B.8) and (B.15), to get successively

$$\int_0^\infty \frac{1-\cos v}{v^{1+d_x+d_y}}\mathrm{d}v = 2\int_0^\infty \frac{\sin^2(v/2)}{v^{1+d_x+d_y}}\mathrm{d}v$$
$$= -\Gamma(-d_x-d_y)\cos(\pi(d_x+d_y)/2)$$
$$= \frac{\pi}{2\Gamma(1+d_x+d_y)\sin(\pi(d_x+d_y)/2)}. \tag{9.37}$$

For the second term of J, (B.20) with $\mu = -d_x - d_y$ and $a = 1$, then (B.8) and (B.15), give

$$\int_0^\infty \frac{\sin v}{v^{1+d_x+d_y}}\mathrm{d}v = \Gamma(-d_x-d_y)\sin(\pi(-d_x-d_y)/2)$$
$$= \frac{\pi}{2\Gamma(1+d_x+d_y)\cos(\pi(d_x+d_y)/2)}. \tag{9.38}$$

Gathering the terms (9.37) and (9.38) under a common denominator and applying (B.1) and (B.8) yields

$$J = \frac{-\mathrm{i}\mathrm{e}^{\mathrm{i}\pi(d_x+d_y)/2}\pi}{\Gamma(1+d_x+d_y)\sin(\pi(d_x+d_y))}.$$

Then, with the further assistance of (B.3), (B.15), and (B.13),

$$\mathrm{e}^{-\mathrm{i}\pi(d_x-d_y)/2}J + \mathrm{e}^{\mathrm{i}\pi(d_x-d_y)/2}\bar{J} = \frac{-\mathrm{i}(\mathrm{e}^{\mathrm{i}\pi d_y} - \mathrm{e}^{-\mathrm{i}\pi d_y})\pi}{\Gamma(1+d_x+d_y)\sin(\pi(d_x+d_y))}$$
$$= \frac{2\pi\sin\pi d_y}{\Gamma(1+d_x+d_y)\sin(\pi(d_x+d_y))}$$
$$= \frac{2\Gamma(1-d_x-d_y)\Gamma(d_x+d_y)\sin\pi d_y}{\Gamma(1+d_x+d_y)}$$
$$= \frac{2\Gamma(1-d_x-d_y)\sin\pi d_y}{d_x+d_y}. \tag{9.39}$$

Finally, on substituting (9.39) into (9.36) the result is (9.34). ∎

As expected, apart from the scale factors $\Gamma(d_x+1)$ and $\Gamma(d_y+1)$ the formula in (9.34) matches λ_{xy} in (4.23). Theorems **9.1**, **9.2**, **9.3**, and **9.4** all reproduce the corresponding formulae obtained for the time domain representation of the fractional process, showing that (2.1) and (9.20) really are alternative representations of the same model. Nonetheless, a comparison of the proofs of Theorems **9.4** and **4.5** is an intriguing exercise, to say the least.

Chapter 10

Autoregressive Roots near Unity

Thinking of the fractional process in its partial-sum manifestation as providing a way of embedding the unit root within a more general class of nonstationary processes, it merits comparison with another model class, that gives rise to Ornstein–Uhlenbeck processes in the limit instead of fractional Brownian motions. What is different about the near-unit root approach is that an array framework is essential. The concept of 'close to unity' is linked to sample size, the limit results being obtained by considering an autoregressive coefficient whose proximity to unity depends on n. Unlike the fractional case, there exist no discrete stationary processes whose normalized partial sums can give rise to the asymptotic limits obtained in this theory. The theory does not so much point to an alternative modelling methodology as to attempt to throw light on the transition between the unit root case and the stable root or mildly explosive root cases of autoregressive models. This chapter is inspired mainly by a seminal paper of Peter Phillips [59]. Related references include among many others [60], [12], [64], and most recently [61].

10.1 Generalizing Unit Roots

Let β be a fixed parameter, define $b_{nj} = \mathrm{e}^{-\beta j/n}$ and let the shock sequence $\{u_i\}_{i=1}^{n}$ satisfy Assumption **1.1**. For each $n \in \mathbb{N}$ consider the triangular moving average array

$$x_{ni} = \sum_{j=0}^{i-1} b_{nj} u_{i-j}, i = 1, \dots, n. \tag{10.1}$$

With $x_{n0} = 0$ the process has the autoregressive representation

$$x_{ni} = \mathrm{e}^{-\beta/n} x_{n,i-1} + u_i, i = 1, \dots, n \tag{10.2}$$

where $\mathrm{e}^{-\beta/n} = 1 - \beta/n + O(n^{-2})$, so that when n is large this is the unit root model with a small-order adjustment.

With $\beta = 0$ this is the true unit root process and $n^{-1/2} x_{n[nt]} \to_{\mathrm{d}} \sigma_u B(t)$ for $t \in [0, 1]$, where B is regular Brownian motion. The case $\beta \neq 0$ models a situation

Asymptotics for Fractional Processes. James Davidson, Oxford University Press. © James Davidson (2025).
DOI: 10.1093/9780198955207.003.0010

in which the autoregressive root, while not unity, is close enough that the process still diverges like $n^{1/2}$. After normalization the limit process, unlike Brownian motion, has dependent increments. With $\beta > 0$, as $n \to \infty$ the distribution is eventually stationary and independent of initial conditions. By contrast, the 'local to unity' fractional process (partial sum of a stationary fractional with parameter $d < \frac{1}{2}$) diverges like $n^{1/2+d}$. After normalization the corresponding limit process (2.1) also has dependent increments but, like Brownian motion, is nonstationary. The three classes of model therefore offer a notable contrast of limit properties.

10.2 The Covariance Function

It will be useful in the sequel to generalize the sequences in (10.1) to the form $\{x_{ni}, i = 1, \dots, [nt]\}$ for any $t > 0$. The case $t > 1$ is important since it allows the asymptotic analysis to extend to continuous time processes defined on the positive half-line. Letting $j = [nt] - i$, the substitution $\mathrm{e}^{-\beta j/n} u_{[nt]-j} = \mathrm{e}^{-\beta([nt]-i)/n} u_i$ is convenient for defining the normalized continuous time càdlàg process

$$X_n(t) = \frac{x_{n[nt]}}{\sqrt{n}} = \frac{\mathrm{e}^{-\beta[nt]/n}}{\sqrt{n}} \sum_{i=1}^{[nt]} \mathrm{e}^{\beta i/n} u_i. \tag{10.3}$$

10.1 Theorem Under Assumption **1.1**,

$$\lim_{n\to\infty} \mathrm{E}(X_n(t)X_n(s)) = \sigma_u^2 \frac{\mathrm{e}^{-\beta|t-s|} - \mathrm{e}^{-\beta(t+s)}}{2\beta}. \tag{10.4}$$

Proof First consider the variance, setting $s = t$. By Assumption **1.1**, $X_n(t)$ is a sum of $[nt]$ independent increments and the variance is

$$\mathrm{E}(X_n(t)^2) = \frac{\sigma_u^2}{n} \mathrm{e}^{-2\beta[nt]/n} \sum_{i=1}^{[nt]} \mathrm{e}^{2\beta i/n}. \tag{10.5}$$

Expressing $\mathrm{e}^{2\beta i/n}$ in power series form as $\sum_{k=0}^{\infty} (2\beta i/n)^k / k!$ and substituting the integral approximation formula

$$\sum_{i=1}^{[nt]} (i/n)^k = \frac{t^{k+1}}{k+1} + O(1/n)$$

for each $k \geq 0$, (10.5) can be written as

$$\begin{aligned} \mathrm{E}(X_n(t)^2) &= \frac{\sigma_u^2}{n} \mathrm{e}^{-2\beta[nt]/n} \sum_{i=1}^{[nt]} \left(\sum_{k=0}^{\infty} \frac{(2\beta i/n)^k}{k!} \right) \\ &= \sigma_u^2 \frac{\mathrm{e}^{-2\beta t}}{2\beta} \sum_{k=0}^{\infty} \frac{(2\beta t)^{k+1}}{(k+1)!} + O(1/n) \\ &= \sigma_u^2 \frac{1 - \mathrm{e}^{-2\beta t}}{2\beta} + O(1/n). \end{aligned} \tag{10.6}$$

In the same way, under Assumption **1.1**,

$$\begin{aligned} \mathrm{E}(X_n(t)X_n(s)) &= \frac{1}{n} \mathrm{e}^{-\beta([nt]+[ns])/n} \sum_{i=1}^{[nt]} \sum_{j=1}^{[ns]} \mathrm{e}^{\beta(i+j)/n} \mathrm{E}(u_i u_j) \\ &= \frac{\sigma_u^2}{n} \mathrm{e}^{-\beta([nt]+[ns])/n} \sum_{i=1}^{[n\min\{t,s\}]} \mathrm{e}^{2\beta i/n} \\ &= \sigma_u^2 \frac{\mathrm{e}^{-\beta(t+s)}}{2\beta} \sum_{k=0}^{\infty} \frac{(2\beta \min\{t,s\})^{k+1}}{(k+1)!} + O(1/n) \\ &= \sigma_u^2 \frac{\mathrm{e}^{-\beta(t+s)}}{2\beta} \left(\mathrm{e}^{2\beta \min\{t,s\}} - 1 \right) + O(1/n). \end{aligned}$$

The limit (10.4) follows, noting that

$$2 \min\{t, s\} - t - s = -|t - s|. \quad \blacksquare \tag{10.7}$$

The limit in (10.4) specifies the covariance function of the Gaussian stochastic process $\sigma_u J_\beta : [0, \infty) \mapsto \mathbb{R}$ where

$$J_\beta(t) = \int_0^t \mathrm{e}^{-\beta(t-\xi)} \mathrm{d}B(\xi) \tag{10.8}$$

and B denotes standard Brownian motion on the positive half-line. Thus,

$$\mathrm{E}(J_\beta(t)J_\beta(s)) = \mathrm{e}^{-\beta(t+s)} \int_0^{\min\{t,s\}} \mathrm{e}^{2\beta\xi} \mathrm{d}\xi = \frac{\mathrm{e}^{-\beta(t+s)}}{2\beta} \left(\mathrm{e}^{2\beta \min\{t,s\}} - 1 \right). \tag{10.9}$$

J_β is the Ornstein-Uhlenbeck process, the solution of the stochastic differential equation

$$\mathrm{d}J_\beta(t) = -\beta J_\beta(t)\mathrm{d}t + \mathrm{d}B(t). \tag{10.10}$$

This can be confirmed by noting that

$$d(e^{\beta\xi}J_\beta(\xi)) = \beta e^{\beta\xi}J_\beta(\xi)d\xi + e^{\beta\xi}dJ_\beta(\xi) = e^{\beta\xi}dB(\xi) \tag{10.11}$$

where the second equality substitutes from (10.10). With $J_\beta(0) = 0$, integrating (10.11) from 0 to t and multiplying through by $e^{-\beta t}$ yields (10.8).

If $\beta > 0$, it is easily seen in view of (10.7) that as $\min\{t, s\} \to \infty$,

$$E(J_\beta(t)J_\beta(s)) \to \frac{e^{-\beta|t-s|}}{2\beta}. \tag{10.12}$$

The limiting case of the variance is therefore simply $1/2\beta$. When t and s are sufficiently large the covariance function depends on $|t - s|$ but not on t or s and the limit process is accordingly stationary. Applying (10.7), it can be verified that $e^{-\beta|t-s|}/2\beta$ is the covariance function of the stationary stochastic process

$$J_\beta^*(t) = \frac{e^{-\beta t}}{\sqrt{2\beta}}B(e^{2\beta t}). \tag{10.13}$$

When $\beta > 0$, $J_\beta(t)$ and $J_\beta^*(t)$ are equivalent Gaussian processes when t is large enough although distinct when t is small, with covariance functions (10.9) and (10.12) respectively.

However, apart from this asymptotic stationarity these results do not depend on the sign of β. When $\beta < 0$ the limit formulae in (10.6) and (10.4) continue to apply, but the variance diverges as $t \to \infty$ instead of converging to the finite limit of $\sigma_u^2/2\beta$. The finite-n autoregression in (10.2) has a mildly explosive root in this case.

10.3 Weak Convergence

The next question to be resolved is the link between (10.8) and the empirical process (10.2). Consider the stochastic process

$$e^{\beta\xi}B(\xi) : [0, t] \mapsto \mathbb{R}.$$

An increment of this product has the form

$$d\left(e^{\beta\xi}B(\xi)\right) = \beta e^{\beta\xi}B(\xi)d\xi + e^{\beta\xi}dB(\xi)$$

and integrating from 0 to t with $B(0) = 0$ yields the integration-by-parts formulation

$$e^{\beta t}B(t) = \beta\int_0^t e^{\beta\xi}B(\xi)d\xi + \int_0^t e^{\beta\xi}dB(\xi). \tag{10.14}$$

Multiplying (10.14) by $e^{-\beta t}$ and rearranging gives for J_β in (10.8) the relation

$$J_\beta(t) = B(t) - \beta\int_0^t e^{-\beta(t-\xi)}B(\xi)d\xi. \tag{10.15}$$

Thus, J_β with $\beta > 0$ can be viewed as the sum of a Brownian motion and a negatively signed bias term based on an average of its recent past variations, tending to push the process in the direction of mean reversion. This characteristic is also evident from the form of (10.10).

To show that $\sigma_u J_\beta$ is indeed the weak limit of X_n in (10.3) consider the element W_n of the space of càdlàg processes on $[0,\infty)$, where

$$W_n(t) = \frac{1}{\sqrt{n}}\sum_{i=1}^{[nt]} u_i. \tag{10.16}$$

Under Assumption **1.1**, Theorem **3.2** with $d = 0$ (for example) is an FCLT for a unit root process that generalizes straightforwardly from the unit interval to the case of $[0, t]$. This gives

$$W_n \xrightarrow{d} \sigma_u B. \tag{10.17}$$

Noting $W_n(0) = 0$ and $W_n(i/n) - W_n((i-1)/n) = n^{-1/2}u_i$, write the telescoping sum representation

$$\begin{aligned} e^{\beta[nt]/n}W_n(t) &= \sum_{i=1}^{[nt]}\left(e^{\beta i/n}W_n(i/n) - e^{\beta(i-1)/n}W_n((i-1)/n)\right) \\ &= \frac{1}{\sqrt{n}}\sum_{i=1}^{[nt]} e^{\beta i/n}u_i + (e^{\beta/n} - 1)\sum_{i=1}^{[nt]} e^{\beta(i-1)/n}W_n((i-1)/n). \end{aligned} \tag{10.18}$$

The first right-hand side sum of (10.18) is $e^{\beta[nt]/n}X_n(t)$ according to (10.3). Considering the second sum, $e^{\beta/n} - 1 = \beta/n + O(1/n^2)$. Since W_n is a step function with $W_n(\xi) = W_n(i/n)$ when $i/n \leq \xi < (i+1)/n$,

$$\frac{1}{n}\sum_{i=1}^{[nt]} e^{\beta(i-1)/n} W_n((i-1)/n) = \sum_{i=0}^{[nt]-1} e^{\beta i/n} W_n(i/n) \int_{i/n}^{(i+1)/n} d\xi$$
$$= \int_0^t e^{\beta\xi} W_n(\xi) d\xi + O_p(1/n). \tag{10.19}$$

Substituting (10.19) into (10.18) and rearranging yields the relation

$$X_n(t) = W_n(t) - \beta \int_0^t e^{-\beta(t-\xi)} W_n(\xi) d\xi + O_p(1/n). \tag{10.20}$$

This is the finite-sample counterpart of (10.15). These arguments strongly indicate that the process $X_n(t)$ has a Gaussian limit in distribution. Since J_β in (10.15) is a continuous functional of B, the following theorem is a consequence of (10.20) and the continuous mapping theorem.

10.2 Theorem If $W_n \to_{\mathrm{d}} \sigma_u B$ then $X_n \to_{\mathrm{d}} \sigma_u J_\beta$. □

However, it is also of interest to pursue direct arguments for the FCLT of the type of used in Theorem **3.2**, since it is evident that the same type of arguments can be applied in both settings. Similarly to the partial sums of fractional processes, X_n is not a simple cumulation of shocks since the moving average coefficients depend on t as well as the shock date i. $X_n(t)$ is a linear function of shocks with nonsummable coefficients so that a CLT can operate, but showing uniform integrability of the increments poses a similar difficulty to that studied in §3.3. The important steps in the following sketch proof have been worked out in detail in Theorem **3.2** and it suffices here to reproduce them briefly.

10.3 Theorem Under Assumption **1.1**, $X_n \to_{\mathrm{d}} \sigma_u J_\beta$.

Proof (Outline) Consider the conditions of Lemma **3.4**. Setting

$$c_{ni} = n^{-1/2} e^{-\beta([nt]-i)/n} = O(n^{-1/2}) \tag{10.21}$$

it is verified in the same way using Theorem **A.5** that

$$\sum_{i=1}^{n} c_{ni}^2 E(u_i^2 1_{\{|c_{ni}u_i|>\varepsilon\}}) = o(n^{1-r/2})$$

for $\varepsilon > 0$, so the Lindeberg condition is satisfied. Under the assumptions this means in view of (10.6) that $X_n(t) \to_d \mathrm{N}(0, \sigma_u^2(1 - \mathrm{e}^{-2\beta t})/2\beta)$ for each $t > 0$. Subject to a.s. continuity, the process X_n/σ_u has a limit with characteristics matching (10.8).

To show uniform tightness, the main requirement is that the collection

$$\left\{ \sup_{t \le s \le t+\delta} \frac{(X_n(s) - X_n(t))^2}{v_n^2(t,\delta)}, n \in \mathbb{N} \right\} \tag{10.22}$$

is uniformly integrable for all $0 < \delta < 1$ and $t \in [0, 1 - \delta]$, where $\sigma_u^2 v_n^2(t,\delta) = \mathrm{E}(X_n(t+\delta) - X_n(t))^2$. Write

$$\begin{aligned} X_n(s) - X_n(t) &= \frac{1}{n^{1/2}} \sum_{i=[nt]+1}^{[ns]} \mathrm{e}^{-\beta([ns]-i)/n} u_i \\ &\quad + \frac{1}{n^{1/2}} \sum_{i=1}^{[nt]} (\mathrm{e}^{-\beta([ns]-i)/n} - \mathrm{e}^{-\beta([nt]-i)/n}) u_i \\ &= Y_{1n}(s,t) + Y_{2n}(s,t). \end{aligned}$$

The squares of these two sums may be considered separately, according to Theorem **A.7**.

In the case of Y_{1n}, direct application of Theorem **A.8** is not possible for the reasons encountered in Lemma **3.6**, that the moving average is not a simple partial sum. In this case, the solution is to consider the partial-sum process

$$Y_{1n}^*(s,t) = \frac{1}{n^{1/2}} \sum_{i=[nt]+1}^{[ns]} \mathrm{e}^{-\beta([n(t+\delta)]-i)/n} u_i.$$

The moving average coefficients $\mathrm{e}^{-\beta([n(t+\delta)]-i)/n}$ for $i = [nt] + 1, \ldots, [n(t+\delta)]$ can be written (with order inverted) as the collection

$$\left\{ 1, \mathrm{e}^{-\beta/n}, \mathrm{e}^{-2\beta/n}, \ldots, \mathrm{e}^{-([n(t+\delta)]-[nt]-1)\beta/n} \right\}. \tag{10.23}$$

This does not depend on s and all the elements appear in the terminal coordinate of the process $Y_{1n}^*(t+\delta, t)$. With

$$v_{1n}^2(t,\delta) = \frac{1}{n} \sum_{i=[nt]+1}^{[n(t+\delta)]} \mathrm{e}^{-2\beta([n(t+\delta)]-i)/n} \tag{10.24}$$

the collection $\{ \sup_{t\le s\le t+\delta} Y^*_{1n}(s,t)^2/v^2_{1n}(t,\delta), n \in \mathbb{N}\}$ satisfies the conditions of Theorem **A.8**. This holds under any permutation p of the moving average weights in (10.23). Let the version of Y^*_{1n} with weights so permuted be denoted Y^*_{1np}. For any n there exists a permutation, say p^*_n, for which

$$\sup_{t\le s\le t+\delta} \frac{Y_{1n}(s,t)^2}{v^2_{1n}(t,\delta)} \le \sup_{t\le s\le t+\delta} \frac{Y^*_{1np^*_n}(s,t)^2}{v^2_{1n}(t,\delta)}. \tag{10.25}$$

The majorants of (10.25) for $n \in \mathbb{N}$ form a uniformly integrable collection by Theorem **A.8**, and this property extends to the collection of the minorants of the inequality, by Theorem **A.6**.

The term $Y_{2n}(s,t)$ is not a partial sum over s and can be dealt with similarly to its counterpart in the proof of Lemma **3.6**. With

$$v^2_{2n}(t,\delta) = \frac{1}{n}\sum_{i=1}^{[nt]}(\mathrm{e}^{-\beta([n(t+\delta)]-i)/n} - \mathrm{e}^{-\beta([nt]-i)/n})^2 \tag{10.26}$$

notice that $v^2_{2n}(t,s-t) \le v^2_{2n}(t,\delta)$ for every s. The collections $\{Y^2_{2n}(s,t)/v^2_{2n}(t,\delta), n \in \mathbb{N}\}$ are uniformly integrable by Corollary **A.9** and Theorem **A.6** for each $s \in [t, t+\delta]$ and hence in particular for the supremum with respect to s.

According to (10.24) $v^2_{1n}(t,\delta) = O(\delta)$, being the sum of $[n(t+\delta)] - [nt]$ positive terms, all below 1 and divided by n. Also, according to (10.26), $v^2_{2n}(t,\delta)$ is a sum of $[nt]$ terms divided by n and the terms have the form

$$(\mathrm{e}^{-\beta([n(t+\delta)]-i)/n} - \mathrm{e}^{-\beta([nt]-i)/n})^2 = O(\delta^2).$$

Since $v^2_n(t,\delta) = v^2_{1n}(t,\delta) + v^2_{2n}(t,\delta)$, the uniform integrability of (10.22) now follows by Theorem **A.7** and the uniform tightness proof is completed by an application of Theorem **3.8** with $L = 0$, $U = 1$, and $d = 0$. ∎

10.4 Stochastic Integral

Applying a telescoping sum argument to the squares of the process leads to another interesting asymptotic relation. Recalling that $x_{ni} = \mathrm{e}^{-\beta i/n}\sum_{k=1}^{i}\mathrm{e}^{\beta k/n}u_k$, let $z_{ni} = \mathrm{e}^{\beta i/n}x_{ni}$ and so note that

$$z^2_{ni} = z^2_{n,i-1} + \mathrm{e}^{2\beta i/n}u^2_i + 2\mathrm{e}^{\beta i/n}z_{n,i-1}u_i. \tag{10.27}$$

Substituting from (10.27) for $z_{ni}^2 - z_{n,i-1}^2$ produces

$$\begin{aligned} x_{ni}^2 - x_{n,i-1}^2 &= \mathrm{e}^{-2\beta i/n} z_{ni}^2 - \mathrm{e}^{-2\beta(i-1)/n} z_{n,i-1}^2 \\ &= \mathrm{e}^{-2\beta i/n}\left(1 - \mathrm{e}^{2\beta/n}\right) z_{ni}^2 + \mathrm{e}^{-2\beta(i-1)/n}\left(z_{ni}^2 - z_{n,i-1}^2\right) \\ &= \mathrm{e}^{-2\beta i/n}\left(1 - \mathrm{e}^{2\beta/n}\right) z_{ni}^2 + \mathrm{e}^{-2\beta(i-1)/n}\left(\mathrm{e}^{2\beta i/n} u_i^2 + 2\mathrm{e}^{\beta i/n} z_{n,i-1} u_i\right) \\ &= \left(1 - \mathrm{e}^{2\beta/n}\right) x_{ni}^2 + \mathrm{e}^{2\beta/n} u_i^2 + 2\mathrm{e}^{\beta/n} x_{n,i-1} u_i. \end{aligned}$$

Given (10.3) and that $x_{n0} = 0$, the telescoping sum therefore has the form

$$\begin{aligned} X_n(1)^2 &= \frac{x_{nn}^2}{n} = \frac{1}{n}\sum_{i=1}^{n}(x_{ni}^2 - x_{n,i-1}^2) \\ &= \left(1 - \mathrm{e}^{2\beta/n}\right)\frac{1}{n}\sum_{i=1}^{n} x_{ni}^2 + \mathrm{e}^{2\beta/n}\frac{1}{n}\sum_{i=1}^{n} u_i^2 + 2\mathrm{e}^{\beta/n}\frac{1}{n}\sum_{i=1}^{n} x_{n,i-1}u_i. \qquad (10.28) \end{aligned}$$

Consider the terms of this equality. $X_n(1)^2 \to_{\mathrm{d}} \sigma_u^2 J_\beta(1)^2$ by Theorem **10.3** and the continuous mapping theorem. Theorem **10.3** also gives

$$\frac{1}{n^2}\sum_{i=1}^{n} x_{ni}^2 = \frac{1}{n}\sum_{i=1}^{n} X_n^2(i/n) = \sum_{i=1}^{n}\int_{(i-1)/n}^{i/n} X_n^2(i/n)\mathrm{d}t \xrightarrow{\mathrm{d}} \sigma_u^2 \int_0^1 J_\beta^2(t)\mathrm{d}t.$$

Next, with W_n defined in (10.16), write $\Delta W_n(i/n) = W_n(i/n+1/n) - W_n(i/n) = n^{-1/2}u_{i+1}$ so that

$$\frac{1}{n}\sum_{i=1}^{n} x_{n,i-1}u_i = \sum_{i=0}^{n-1} X_n(i/n)\Delta W_n(i/n) \xrightarrow{\mathrm{d}} \sigma_u \int_0^1 J_\beta(t)\mathrm{d}B(t)$$

where the indicated limit is an Itô integral with respect to Brownian motion, having zero mean. Finally, applying the law of large numbers to the sequence u_i^2 and noting $\mathrm{e}^{\beta/n} \to 1$ and $n(1 - \mathrm{e}^{2\beta/n}) \to -2\beta$, after dividing by σ_u^2 and rearrangement, as $n \to \infty$ the almost sure limiting form of the relation in (10.28) is

$$\int_0^1 J_\beta(t)\mathrm{d}B(t) = \tfrac{1}{2}\left(J_\beta^2(1) - 1\right) + \beta\int_0^1 J_\beta^2(t)\mathrm{d}t. \qquad (10.29)$$

In one sense, this expression might be viewed as generalizing to $\beta \neq 0$ the well-known property of Brownian motions, that $\int_0^1 B\mathrm{d}B = \frac{1}{2}(B(1)^2 - 1)$ with probability 1. However, the true generalization has the form

$$\int_0^1 J_\beta \mathrm{d}J_\beta = \tfrac{1}{2}(J_\beta^2(1) - 1). \tag{10.30}$$

To verify this claim consider the increment of (10.3),

$$\begin{aligned}\Delta X_n(t) &= \frac{\mathrm{e}^{-\beta([nt]+1)/n}}{n^{1/2}} \sum_{i=1}^{[nt]+1} \mathrm{e}^{\beta i/n} u_i - \frac{\mathrm{e}^{-\beta[nt]/n}}{n^{1/2}} \sum_{i=1}^{[nt]} \mathrm{e}^{\beta i/n} u_i \\ &= \frac{u_{[nt]+1}}{n^{1/2}} + (\mathrm{e}^{-\beta/n} - 1)\frac{\mathrm{e}^{-\beta[nt]/n}}{n^{1/2}} \sum_{i=1}^{[nt]} \mathrm{e}^{\beta i/n} u_i \\ &= \Delta W_n(t) - \beta\frac{1}{n}X_n(t) + O_p(n^{-2}). \end{aligned} \tag{10.31}$$

Multiply (10.31) by $X_n(t)$ to give

$$X_n(t)\Delta X_n(t) = X_n(t)\Delta W_n(t) - \beta\frac{1}{n}X_n^2(t) + O_p(n^{-2})$$

and then integrate with respect to t over $[0, 1]$. The normalized limiting relation as $n \to \infty$ has the form

$$\int_0^1 J_\beta(t)\mathrm{d}J_\beta(t) = \int_0^1 J_\beta(t)\mathrm{d}B(t) - \beta \int_0^1 J_\beta^2(t)\mathrm{d}t \tag{10.32}$$

where the right-hand side of the equation can be taken as defining the left-hand side. This calculation shows that (10.32) is not an Itô integral and has a nonzero mean whose sign depends on β. That this is negative when $\beta > 0$ is a further indication of the mean-reversion tendency of the Ornstein–Uhlenbeck process.

10.5 Autocorrelated Shocks

This chapter has so far invoked Assumption **1.1** on the usual grounds of clarity and simplicity, but it would not be difficult to extend the weak convergence argument to Assumption **1.2**, using Lemma **3.13** as the model to modify Theorem **10.3**. The conditions of Assumption **3.9** are easily verified for the c_{ni} in (10.21).

However, extending Theorem **10.1** to autocorrelated shock processes is complicated in much the same way that going from Corollary **2.8** to Theorem **2.10** proved nontrivial. The following result extends Theorem **10.1** to weakly dependent shocks, although only for the case of the variance. The counterpart of (10.4) with the replacement of σ_u^2 by ω_u^2 can be obtained in the same way with some further notational overhead.

10.4 Theorem If Assumption **1.2** holds for u_i in (10.3) then

$$\lim_{n\to\infty} \mathrm{E}(X_n(t)^2) = \omega_u^2 \frac{1 - \mathrm{e}^{-2\beta t}}{2\beta}, \; t \geq 0.$$

Proof Choose a monotone integer sequence $\{B_n \in \mathbb{N}\}$ such that $B_n \to \infty$ but $B_n/n \to 0$ as $n \to \infty$, and put $r_n = [nt/B_n]$ whenever n is large enough that $r_n \geq 1$. Assume for convenience that t takes a value such that $[nt] = r_n B_n$. This is harmless since by taking n large enough, every t can be made arbitrarily close to a value obeying the restriction.

Then, (10.3) can be written as

$$X_n(t) = \mathrm{e}^{-\beta[nt]/n} \frac{[nt]^{1/2}}{n^{1/2}} \frac{1}{r_n^{1/2}} \sum_{j=1}^{r_n} \left(\frac{1}{B_n^{1/2}} \sum_{i=(j-1)B_n+1}^{jB_n} \mathrm{e}^{\beta i/n} u_i \right). \tag{10.33}$$

Define

$$a_{nji} = \frac{\mathrm{e}^{\beta i/n} - \mathrm{e}^{\beta j B_n/n}}{\mathrm{e}^{\beta j B_n/n}} = \mathrm{e}^{\beta(i - jB_n)/n} - 1$$

and

$$a_n^* = \mathrm{e}^{-\beta B_n/n} - 1. \tag{10.34}$$

The r_n bracketed terms in (10.33) can each be decomposed as

$$\frac{1}{B_n^{1/2}} \sum_{i=(j-1)B_n+1}^{jB_n} \mathrm{e}^{\beta i/n} u_i = \mathrm{e}^{\beta j B_n/n} S_{nj} + \mathrm{e}^{\beta j B_n/n} a_n^* S_{nj}^* \tag{10.35}$$

where

$$S_{nj} = \frac{1}{B_n^{1/2}} \sum_{i=(j-1)B_n+1}^{jB_n} u_i$$

and

$$S_{nj}^* = \frac{1}{B_n^{1/2}} \sum_{i=(j-1)B_n+1}^{jB_n} \frac{a_{nji}}{a_n^*} u_i.$$

The signs of a_{nji} and a_n^* depend on that of β, but always match since $i \leq jB_n$, so their ratio is positive in every case and $0 \leq a_{nji}/a_n^* \leq 1$.

With these definitions, consider $\mathrm{E}(X_n^2(t))$. Multiplying out the expected square of (10.33) after substituting (10.35), there are three types of summand: the expected squares and products of the S_{nj} (r_n^2 terms), the expected squares and products of the S_{nj}^* (r_n^2 terms), and the expected products of the S_{nj}^* and S_{nj} ($2r_n^2$ terms). The terms of the first type have the form

$$\mathrm{E}\left(\frac{1}{r_n^{1/2}}\sum_{j=1}^{r_n} \mathrm{e}^{\beta j B_n/n} S_{nj}\right)^2 = T_{1n} + 2T_{2n}$$

where

$$T_{1n} = \frac{1}{r_n}\sum_{j=1}^{r_n} \mathrm{e}^{2\beta j B_n/n}\mathrm{E}(S_{nj}^2)$$

and

$$T_{2n} = \frac{1}{r_n}\sum_{j=2}^{r_n} \mathrm{e}^{\beta j B_n/n} \sum_{m=1}^{j-1} \mathrm{e}^{\beta(j-m)B_n/n}\mathrm{E}(S_{nj}S_{n,j-m}).$$

Since $\mathrm{E}(S_{nj}^2) \to \omega_u^2$ as $n \to \infty$ for every j,

$$T_{1n} \to \omega_u^2 \int_0^1 \mathrm{e}^{2\beta t x}\mathrm{d}x = \omega_u^2 \frac{\mathrm{e}^{2\beta t} - 1}{2\beta t}. \tag{10.36}$$

Also, in view of the fact that $|\mathrm{E}(S_{nj}S_{n,j-m})| = O(m^{-1-\delta}B_n^{-\delta})$ for all j under Assumption **1.2** (compare the calculation in (2.61)), $T_{2n} = O(B_n^{-\delta})$.

Next, define T_{1n}^* and T_{2n}^* to have the same form as T_{1n} and T_{2n} except that S_{nj} is replaced by S_{nj}^*. $\mathrm{E}(S_{nj}^2)$ and $\mathrm{E}(S_{nj}^{*2})$ are quadratic forms with respect to the same matrix of expected products, with weights that are unity in the former case and drawn from the interval $[0, 1]$ in the latter case. Since $\mathrm{E}(S_{nj}^2) = O(1)$, the same is true of $\mathrm{E}(S_{nj}^{*2})$ and by the same reasoning as before, $T_{1n}^* = O(1)$ and $T_{2n}^* = O(B_n^{-\delta})$. The terms of the second type therefore have the form

$$\mathrm{E}\left(\frac{1}{r_n^{1/2}}\sum_{j=1}^{r_n} \mathrm{e}^{\beta j B_n/n} a_n^* S_{nj}^*\right)^2 = a_n^{*2}(T_{1n}^* + 2T_{2n}^*) = O(r_n^{-2}) \tag{10.37}$$

where the indicated order of magnitude of (10.37) follows since by (10.34), $|a_n^*| = |\mathrm{e}^{-\beta/r_n} - 1| = O(r_n^{-1})$.

By the same reasoning, the terms of the third type have the form

$$a_n^* \frac{2}{r_n} \mathrm{E}\left(\sum_{j=1}^{r_n} e^{\beta j B_n/n} S_{nj} \sum_{j'=1}^{r_n} e^{\beta j' B_n/n} S_{nj'}^* \right) = O(r_n^{-1}).$$

The conclusion according to (10.33) is that

$$\mathrm{E}(X_n(t)^2) = e^{-2\beta[nt]/n} \frac{[nt]}{n} T_{1n} + O(\max\{B_n^{-\delta}, r_n^{-1}\}). \tag{10.38}$$

The theorem follows on putting together (10.38) and (10.36). ∎

Appendix A: Useful Results

This appendix proves some results that do not relate explicitly to the fractional model, but are useful in particular for establishing various steps in the proof of the fractional FCLT.

A.1 Theorem Let the function $L : [0, \infty) \mapsto \mathbb{R}$ be slowly varying at ∞.

(i) If $\rho > -1$,

$$\int_0^v y^\rho L(y)\mathrm{d}y \sim \frac{v^{1+\rho}L(v)}{1+\rho} \text{ as } v \to \infty$$

(ii) If $\rho < -1$

$$\int_v^\infty y^\rho L(y)\mathrm{d}y \sim -\frac{v^{1+\rho}L(v)}{1+\rho} \text{ as } v \to \infty.$$

Proof For (i), make the change of variable $y = sv$ where $0 \leq s \leq 1$ and write

$$\int_0^v y^\rho L(y)\mathrm{d}y = v\int_0^1 (sv)^\rho L(sv)\mathrm{d}s = v^{1+\rho}L(v)\int_0^1 s^\rho \frac{L(sv)}{L(v)}\mathrm{d}s.$$

For (ii), make the change of variable $y = sv$ where $1 \leq s < \infty$ and write

$$\int_v^\infty y^\rho L(y)\mathrm{d}y = v\int_1^\infty (sv)^\rho L(sv)\mathrm{d}s = v^{1+\rho}L(v)\int_1^\infty s^\rho \frac{L(sv)}{L(v)}\mathrm{d}s.$$

In each case, the theorem follows since $L(sv)/L(v) \to 1$ as $v \to \infty$. ∎

In part (ii), the case $\rho = -1$ can have a solution depending on the form of L. If $L(y) = \log(y)^{-1-a}$ for $a > 0$, make the change of variable $u = \log(y)$ so that

$\mathrm{d}u = y^{-1}\mathrm{d}y$. Then for $v > 1$,

$$\int_v^\infty y^{-1}\log(y)^{-1-a}\mathrm{d}y = \left[\frac{1}{-a}\log(y)^{-a}\right]_v^\infty = \frac{1}{a\log(v)^a}. \tag{A.1}$$

Here are two well-known identities relating to a random variable X and constants $a > 0$ and $x > 0$.[1]

A.2 Theorem $\mathrm{E}(|X|^a 1_{\{|X|\le x\}}) = a\int_0^x \xi^{a-1}P(|X| > \xi)\mathrm{d}\xi - x^a P(|X| > x)$. □

A.3 Corollary $\mathrm{E}(|X|^a) = a\int_0^\infty \xi^{a-1}P(|X| > \xi)\mathrm{d}\xi$. □

The first follows easily, applying integration by parts after writing $P(|X| > x) = 1 - F(x)$ where F is the c.d.f. of $|X|$. The second one shows the case $x = \infty$ making use of $x^a = a\int_0^x \xi^{a-1}\mathrm{d}\xi$.

A.4 Lemma For a random variable X, L_r-boundedness implies

$$P(|X| > \eta) = O(\eta^{-r}\log(\eta)^{-1-\mu}) \tag{A.2}$$

for $\mu > 0$.

Proof Corollary **A.3** with $a = 1$ gives the equality $\mathrm{E}(|X|) = \int_0^\infty P(|X| > \xi)\mathrm{d}\xi$. Applied to the case $|X|^r$, this means that

$$\mathrm{E}(|X|^r) = \int_0^\infty P(|X|^r > \xi)\mathrm{d}\xi. \tag{A.3}$$

L_r-boundedness implies integrability on the right hand side so that

$$P(|X|^r > \xi) = O(\xi^{-1}\log(\xi)^{-1-\mu})$$

as $\xi \to \infty$, from which (A.2) follows on setting $\eta = \xi^{1/r}$. ∎

[1] These are proved in SLT as Theorem 9.21 and Corollary 9.22, respectively.

A.5 Theorem If a random variable X is L_r-bounded for $r \geq 2$ then as $\eta \to \infty$,

$$\mathrm{E}(X^2 1_{\{|X|>\eta\}}) = o(\eta^{2-r}). \tag{A.4}$$

Proof Begin by writing

$$\begin{aligned}\mathrm{E}(X^2 1_{\{|X|>\eta\}}) &= \mathrm{E}(X^2) - \mathrm{E}(X^2 1_{\{|X|\leq\eta\}}) \\ &= 2\int_\eta^\infty \xi P(|X| > \xi)\mathrm{d}\xi + \eta^2 P(|X| > \eta) \end{aligned} \tag{A.5}$$

where the second equality is got by applying Corollary **A.3** and Theorem **A.2** with $a = 2$.

For clarity the cases $r > 2$ and $r = 2$ are treated separately. In the first case, Lemma **A.4** implies that $\eta^r P(|X| > \eta) = O(\log(\eta)^{-1-\mu})$ as $\eta \to \infty$. Hence, when η is large enough, $\xi > \eta$ implies

$$\xi^r P(|X| > \xi) \leq \eta^r P(|X| > \eta). \tag{A.6}$$

Multiplying through (A.6) by ξ^{1-r} and integrating both sides from η to ∞ produces the inequality

$$\begin{aligned}\int_\eta^\infty \xi P(|X| > \xi)\mathrm{d}\xi &\leq \eta^r P(|X| > \eta)\int_\eta^\infty \xi^{1-r}\mathrm{d}\xi \\ &= \frac{1}{r-2}\eta^2 P(|X| > \eta). \end{aligned} \tag{A.7}$$

Substituting (A.7) into (A.5) gives

$$\mathrm{E}(X^2 1_{\{|X|>\eta\}}) \leq \frac{r}{r-2}\eta^2 P(|X| > \eta).$$

Finally, apply (A.2) to get (A.4).

For the case $r = 2$, first set $\nu = \mu/2$ so that Lemma **A.4** gives

$$\eta^2 \log(\eta)^{1+\nu} P(|X| > \eta) = O(\log(\eta)^{-\nu}). \tag{A.8}$$

Therefore, $\xi > \eta$ implies that

$$\xi^2 \log(\xi)^{1+\nu} P(|X| > \xi) \leq \eta^2 \log(\eta)^{1+\nu} P(|X| > \eta)$$

when η is large enough. Multiply both sides of this inequality by $\xi^{-1}\log(\xi)^{-1-\nu}$ and integrate from η to ∞ to get

$$\int_\eta^\infty \xi P(|X| > \xi)\mathrm{d}\xi \le \eta^2 \log(\eta)^{1+\nu} P(|X| > \eta) \int_\eta^\infty \xi^{-1}\log(\xi)^{-1-\nu}\mathrm{d}\xi. \quad \text{(A.9)}$$

The right-hand-side integral has the solution $1/(\nu\log(\eta)^\nu)$ for $\eta > 1$, as shown in (A.1). Substituting from (A.9) into (A.5) gives, according to (A.8),

$$\begin{aligned}\mathrm{E}(X^2 1_{\{|X|>\eta\}}) &\le \eta^2\left(\frac{2\log(\eta)}{\nu} + 1\right)P(|X| > \eta)\\ &= O(\log(\eta)^{-\mu}) = o(1)\end{aligned}$$

as $\eta \to \infty$ which matches (A.4) for this case. ∎

A.6 Theorem If U and V are L_2-bounded random variables and $|U| \le |V|$ a.s., then for a constant $M > 0$,

$$\mathrm{E}(U^2 1_{\{|U|>M\}}) \le \mathrm{E}(V^2 1_{\{|V|>M\}}).$$

Proof The assumption implies $U^2 1_{\{|U|>M\}} \le V^2 1_{\{|V|>M\}}$ a.s. and $X \le Y$ a.s. implies that $\mathrm{E}(X) \le \mathrm{E}(Y)$ for any pair of integrable random variables X and Y. ∎

A.7 Theorem If X and Y are L_2-bounded random variables,

$$\mathrm{E}((X+Y)^2 1_{\{|X+Y|>M\}}) \le 4\max\{\mathrm{E}(X^2 1_{\{|X|>M/2\}}), \mathrm{E}(Y^2 1_{\{|Y|>M/2\}})\}$$

for $M > 0$.

Proof For nonnegative L_2-bounded random variables U and V,

$$\begin{aligned}&\mathrm{E}\left((U+V)^2 1_{\{U+V>M\}}\right)\\ &\quad= \mathrm{E}\left((U+V)^2 1_{\{U+V>M\}\cap\{U\ge V\}}\right) + \mathrm{E}\left((U+V)^2 1_{\{U+V>M\}\cap\{U<V\}}\right)\\ &\quad\le \max\left\{\mathrm{E}\left((2U)^2 1_{\{2U>M\}}\right), \mathrm{E}\left((2V)^2 1_{\{2V>M\}}\right)\right\}\\ &\quad= 4\max\left\{\mathrm{E}\left(U^2 1_{\{U>M/2\}}\right), \mathrm{E}\left(V^2 1_{\{V>M/2\}}\right)\right\} \qquad \text{(A.10)}\end{aligned}$$

where the inequality follows by two applications of Theorem **A.6**. Setting $U = |X|$ and $V = |Y|$, $|X+Y| \le |X| + |Y|$ by the triangle inequality and

$$\mathrm{E}((X+Y)^2 1_{\{|X+Y|>M\}}) \le \mathrm{E}((|X|+|Y|)^2 1_{\{|X|+|Y|>M\}}) \qquad \text{(A.11)}$$

by a further application of Theorem **A.6**, which in conjunction with (A.10) completes the proof. ∎

The next result is a leading implication of **A.5**, strengthening the simple demonstration of uniform integrability by specifying the rate of convergence, which is important for certain applications.

A.8 Theorem Let $\{u_i, -\infty < i < \infty\}$ be independently distributed and uniformly L_r-bounded for $r \geq 2$, and let $S_k = \sum_{j=1}^{k} c_{nj}u_j$ where $\{c_{nj}\}$ is a triangular array of constants and $\sum_{j=1}^{n} c_{nj}^2 = O(1)$ as $n \to \infty$. The collection $\{\max_{1\leq k\leq n} S_k^2, n \in \mathbb{N}\}$ is uniformly integrable with

$$\mathrm{E}\left(\max_{1\leq k\leq n} S_k^2 1_{\{\max_{1\leq k\leq n}|S_k|>\eta\}}\right) = o(\eta^{2-r}) \tag{A.12}$$

as $\eta \to \infty$.

Proof For finite constants $K_1 > 0, K_2 > 0$, and $K_3 \geq K_2 \sum_{j=1}^{n} c_{nj}^2$ where $K_3 < \infty$ exists by assumption,

$$\begin{aligned}\mathrm{E}\left(\max_{1\leq k\leq n}|S_k|^r\right) &\leq K_1\mathrm{E}|S_n|^r \leq K_2\mathrm{E}\left(\sum_{j=1}^{n} c_{nj}^2u_j^2\right)^{r/2} \\ &\leq K_2\sum_{j=1}^{n} c_{nj}^2\mathrm{E}|u_j|^r \leq K_3\max_{1\leq j\leq n}\mathrm{E}|u_j|^r.\end{aligned} \tag{A.13}$$

In (A.13) the first inequality is the Doob inequality,[2] the second the Burkholder inequality,[3] and the third is by convexity. Noting that

$$\max_{1\leq k\leq n}|S_k|^r = \left(\max_{1\leq k\leq n}|S_k|\right)^r$$

and that the random variable $\max_{1\leq k\leq n}|S_k|$ is L_r-bounded by (A.13) and the assumption on $\{u_j\}$, it follows by Theorem **A.5** that (A.12) holds as $\eta \to \infty$. Given the assumptions, this conclusion does not depend on n and so holds for $n \in \mathbb{N}$. ∎

This result holds equally in the case where $\{u_i\}$ is a martingale difference, although that extension is not used in the present applications. Both the Doob

[2] SLT Theorem 16.21.
[3] SLT Theorem 16.24.

and Burkholder inequalities are martingale results, therefore holding under independence, although the somewhat simpler Marcinkiewicz–Zygmund inequality[4] can substitute for Burkholder's in that case.

A.9 Corollary Under the conditions of Theorem **A.8**, the collection $\{S_n^2, n \in \mathbb{N}\}$ is uniformly integrable with

$$\mathrm{E}\left(S_n^2 1_{\{|S_n|>\eta\}}\right) = o(\eta^{2-r}). \tag{A.14}$$

Proof Immediate by Theorem **A.6**, since $S_n^2 \leq \max_{1\leq k\leq n} S_k^2$. ∎

[4] For example, see [33] Theorem 8.1.

Appendix B: Identities and Integral Solutions

$$e^{ix} = \cos x + i \sin x \tag{B.1}$$

$$\cos x = \frac{e^{ix} + e^{-ix}}{2} \tag{B.2}$$

$$\sin x = \frac{e^{ix} - e^{-ix}}{2i} \tag{B.3}$$

$$\cos(x \pm \pi) = -\cos x \tag{B.4}$$

$$\sin(x \pm \pi) = -\sin x \tag{B.5}$$

$$\cos\left(\frac{\pi}{2} - x\right) = \sin x \tag{B.6}$$

$$\sin\left(\frac{\pi}{2} - x\right) = \cos x \tag{B.7}$$

$$\sin(2x) = 2\cos x \sin x \tag{B.8}$$

$$\sin^2 x = \frac{1 - \cos(2x)}{2} \tag{B.9}$$

$$\sin x + \sin y = 2\cos\left(\frac{x - y}{2}\right)\sin\left(\frac{x + y}{2}\right) \tag{B.10}$$

$$\cos(x - y) - \cos(x + y) = 2\sin x \sin y \tag{B.11}$$

$$\Gamma(x) = \int_0^{\infty} t^{x-1} e^{-t} dt \ (x > 0) \tag{B.12}$$

$$\Gamma(x + 1) = x\Gamma(x) \ (x \neq 0, -1, -2, \ldots) \tag{B.13}$$

$$B(x, y) = \int_0^1 t^{x-1}(1 - t)^{y-1} dt = \frac{\Gamma(x)\Gamma(y)}{\Gamma(x + y)} \ (x > 0, y > 0) \tag{B.14}$$

$$\text{Euler reflection formula: } \Gamma(x)\Gamma(1-x)\sin(\pi x) = \pi \; (x \notin \mathbb{Z}) \tag{B.15}$$

$$\text{Stirling's approximation: } \log(n!) = n\log n - n + O(\log n) \tag{B.16}$$

$$\int_{-\pi}^{\pi} e^{ij\lambda} d\lambda = \begin{cases} 2\pi & j = 0 \\ 0, & j \neq 0, \text{ integer} \end{cases} \tag{B.17}$$

$$\int_{0}^{\pi} \cos kx \cos jx dx = \begin{cases} \pi/2 & j = k \neq 0 \\ 0 & j \neq k \end{cases} \quad j, k \text{ integers} \tag{B.18}$$

$$\int_{0}^{\pi} \sin^{\nu-1} x \cos(ax) dx = \frac{\pi}{2^{\nu-1}\nu B\left(\frac{\nu+a+1}{2}, \frac{\nu-a+1}{2}\right)} \cos\left(\frac{a\pi}{2}\right) \tag{B.19}$$

$(\nu > 0)$ [30], eqn. 3.631.8.

$$\int_{0}^{\infty} x^{\mu-1} \sin(ax) dx = \frac{\Gamma(\mu)}{a^{\mu}} \sin\left(\frac{\mu\pi}{2}\right) \tag{B.20}$$

$(a > 0, 0 < |\mu| < 1)$ [30], eqn. 3.761.4.

$$\int_{0}^{\infty} x^{\mu-1} \sin^2(ax) dx = -\frac{\Gamma(\mu)}{2^{\mu+1}a^{\mu}} \cos\left(\frac{\mu\pi}{2}\right) \tag{B.21}$$

$(a > 0, -2 < \mu < 0)$ [30], eqn. 3.823.

$$F(a, b; c; z) = \frac{\Gamma(c)}{\Gamma(a)\Gamma(b)} \sum_{j=0}^{\infty} \frac{\Gamma(a+j)\Gamma(b+j)}{\Gamma(c+j)\Gamma(j+1)} z^j \tag{B.22}$$

$(|z| \leq 1, c \neq 0, -1, -2, \ldots)$ [1], eqn. 15.1.1.

$$F(a, b; c; 1) = \frac{\Gamma(c)\Gamma(c-a-b)}{\Gamma(c-a)\Gamma(c-b)} \tag{B.23}$$

$(c - a - b > 0)$ [1], eqn. 15.1.20.

$$F(a, b; c; z) = \frac{\Gamma(c)}{\Gamma(b)\Gamma(c-b)} \int_{0}^{1} \tau^{b-1}(1-\tau)^{c-b-1}(1-\tau z)^{-a} d\tau \tag{B.24}$$

$(c > b > 0)$ [1], eqn. 15.3.1.

References

[1] Abramowitz, M. and I. A. Stegun (1965) *Handbook of Mathematical Functions.* Dover, New York.

[2] Abadir, K., W. Distaso, L. Giraitis, and H. L. Koul (2014) Asymptotic normality for weighted sums of linear processes, *Econometric Theory* 30, 252–284.

[3] Banerjee, A., J. J. Dolado, J. W. Galbraith, and D. F. Hendry (1993) *Co-integration, Error Correction, and the Econometric Analysis of Non-Stationary Data,* Advanced Texts in Econometrics. Oxford University Press, Oxford.

[4] Beran, J. (1994) *Statistics for Long Memory Processes.* Chapman and Hall, New York.

[5] Beran, J., Y. Feng, S. Ghosh, and R. Kulik (2013) *Long-Memory Processes, Probabilistic Properties and Statistical Methods* Springer, Heidelberg.

[6] Billingsley, P. (1968) *Convergence of Probability Measures,* John Wiley and Sons, New York.

[7] Billingsley, P. (1999) *Convergence of Probability Measures* (2nd Edn.), John Wiley and Sons, New York.

[8] Bondon, P. and W. Palma (2007) A class of antipersistent processes. *Journal of Time Series Analysis* 28, 261–273.

[9] Box, G. E. P., G. M. Jenkins, G. C. Reinsel, and G. M. Ljung (2016) *Time Series Analysis: Forecasting and Control* (5th Edn.), John Wiley & Sons, Hoboken NJ.

[10] Brockwell, P. J. and R. A. Davis (1991) *Time Series: Theory and Methods,* (2nd Edn.), Springer-Verlag, New York.

[11] Chan, N. H. and N. Terrin (1995) Inference for unstable long memory processes with applications to fractional unit root autoregressions. *Annals of Statistics* 23, 1662–1683.

[12] Chan, N. H. and C.-Z. Wei (1987) Asymptotic inference for nearly nonstationary AR(1) processes, *Annals of Statistics* 15, 1050–1063.

[13] Davidson, J. (2000) *Econometric Theory,* Blackwell Publishers, Oxford.

[14] Davidson, J. (2021) *Stochastic Limit Theory: An Introduction for Econometricians,* (2nd Edn.), Oxford University Press, Oxford.

[15] Davidson, J. (2002) Establishing conditions for the functional central limit theorem in nonlinear and semiparametric time series processes, *Journal of Econometrics* 106, 243–269.

[16] Davidson, J. (2004) Moment and memory properties of linear conditional heteroscedasticity models, and a new model. *Journal of Business and Economics Statistics* 22 (1), 16–29.

[17] Davidson, J. and R. M. de Jong (2000) The functional central limit theorem and convergence to stochastic integrals II: fractionally integrated processes. *Econometric Theory* 16(5), 643–666.

[18] Davidson, J. and N. Hashimzade (2008) Alternative Frequency and Time Domain Versions of Fractional Brownian Motion, *Econometric Theory* 24(1), 256–293.

[19] Davidson, J. and N. Hashimzade (2009a) Type I and type II fractional Brownian motions: a reconsideration *Computational Statistics and Data Analysis* 53(6) (2009) 2089–2106.

[20] Davidson, J. and N. Hashimzade (2009b) Representation and Weak Convergence of Stochastic Integrals with Fractional Integrator Processes", *Econometric Theory* 25(6), 1589–1624.

[21] Davydov, Yu. A. (1970) The invariance principle for stationary processes. *Theory of Probability and its Applications* 15(3), 487–498.
[22] De Jong, R. M. (1997) Central limit theorems for dependent heterogeneous random variables. *Econometric Theory* 13(3) 353–367.
[23] De Jong, R. M. and Davidson, J. (2000) The functional central limit theorem and convergence to stochastic integrals I: weakly dependent processes. *Econometric Theory* 16(5), 621–642.
[24] Doukhan, P., G. Oppenheim, and M. Taqqu (eds.) (2003) *Theory and applications of long-range dependence.* Birkhauser, Boston.
[25] Duncan, T. E., Y. Hu, and B. Pasik-Duncan (2000) Stochastic calculus for fractional Brownian motion. I: Theory. *SIAM Journal of Control Optimization* 38, 582–612.
[26] Fishman, G. S. (1969) *Spectral Methods in Econometrics.* Harvard University Press, Cambridge Mass.
[27] Geweke, J. and S. Porter-Hudak (1983) The estimation and application of long memory time series models, *Journal of Time Series Analysis* 4, 221–238.
[28] Giraitis, L., H. L. Koul, and D. Surgailis (2012) *Large Sample Inference for Long Memory Processes.*, Imperial College Press, London.
[29] Gorodetskii, V. V. (1977) On convergence to semi-stable Gaussian processes, *Theory of Probability and its Applications* 22(3), 498–508.
[30] Gradsteyn, I. S. and I. M. Ryzhik (2007) *Table of Integrals, Series and Products* (7th Edn.) Academic Press, Burlington, MA.
[31] Granger, C. W. J. (1980) 'Long memory relationships and the aggregation of dynamic models', *Journal of Econometrics* 14, 227–238.
[32] Granger, C. W. J. and R. Joyeux (1980) An introduction to long memory time series models and fractional differencing. *Journal of Time Series Analysis* 1(1) 15–29.
[33] Gut, A. (2013), *Probability: a Graduate Course* (2nd Edn.), Springer, New York.
[34] Hassler, U. (2019) *Time Series Analysis with Long Memory in View*, John Wiley & Sons, Hoboken, NJ.
[35] Hurvich, C. M. and K. I. Beltrao (1994) Automatic semiparametric estimation of the memory parameter of a long memory time series, *Journal of Time Series Analysis* 15, 285–302.
[36] Hosking, J. R. M. (1981) Fractional differencing. *Biometrika* 68(1), 165–176.
[37] Hosking, J. R. M. (1984a) Modelling persistence in hydrological time series using fractional differencing, *Water Resources Research* 20(12), 1898–1908.
[38] Hosking, J. R. M. (1984b) Asymptotic distributions of the sample mean, autocovariances and autocorrelations of long memory time series. Technical summary report 2752, Mathematics Research Center, University of Wisconsin, Madison WI.
[39] Hosking, J. R. M. (1996) Asymptotic distributions of the sample mean, autocovariances and autocorrelations of long memory time series, *Journal of Econometrics* 73, 261–284.
[40] Hosoya, Y. (2003) Fractional invariance principle, *Journal of Time Series Analysis* 26(3), 463–486.
[41] Hurst, H. E. (1951) Long term storage capacity of reservoirs, *Transactions of the American Society of Civil Engineers* 116, 770–779.
[42] Johansen S. and M. Ø. Nielsen (2011) A necessary moment condition for the fractional functional central limit theorem, *Econometric Theory* 28(3), 671–679.
[43] Lamperti, J. (1962) Semi-stable stochastic processes. *Transactions of the American Mathematical Society* 104(1), 62–78.
[44] Karatsas, I. and S. Shreve (1991) *Brownian Motion and Stochastic Calculus*, (2nd Edn.), Springer-Verlag, New York.

[45] Liu, M. (1998) Asymptotics of nonstationary fractional integrated series, *Econometric Theory* 14, 641–662.
[46] Lo, A. W. (1991) Long-term memory in stock market prices, *Econometrica* 59(5), 1279–1313.
[47] Mandelbrot, B. B. and J. R. Wallis (1968) Noah, Joseph and operational hydrology, *Water Resources Research* 4(5), 909–918.
[48] Mandelbrot, B. B. and J. W. van Ness 1968. Fractional Brownian motions, fractional noises and applications, SIAM Review 10, 4, 422–437.
[49] Marinucci, D. and P. M. Robinson (1999) Weak convergence to fractional Brownian motion, *Stochastic Processes and their Applications* 80, 103–120.
[50] Marinucci, D. and P. M. Robinson (1999) Alternative forms of fractional Brownian motion, *Journal of Statistical Inference and Planning* 80, 111–122.
[51] Mason. D. M. (2016) The Hurst phenomenon and the rescaled range statistic, *Stochastic Processes and their Applications* 126, 3790–3807.
[52] McLeish, D. L. (1975), Invariance principles for dependent variables, *Z. Wahrscheinlichkeitstheorie verw. Gebiete* 32, 165–178.
[53] McLeish, D. L. (1977), On the invariance principle for nonstationary mixingales, *Annals of Probability* 5(4), 616–621.
[54] Moulines, E. and P. Soulier (1999) Broad band log-periodogram estimation of time series with long-range dependence, *Annals of Statistics* 27, 1415–1439.
[55] Odaki, M. (1993) On the invertibility of fractionally differenced ARIMA processes, *Biometrika* 80(3), 703–709.
[56] Palma, W. (2007) *Long Memory Time Series: Theory and Methods*, John Wiley & Sons, Hoboken NJ.
[57] Palma, W. (2016) *Time Series Analysis*, John Wiley & Sons, Hoboken NJ.
[58] Pipiras, V. and M. Taqqu (2017) *Long-Range Dependence and Self-Similarity*, Cambridge University Press, Cambridge.
[59] Phillips, P. C. B. (1987) Towards a unified asymptotic theory for autoregression, *Biometrika* 74, 535–547.
[60] Phillips, P. C. B. (1988) Regression theory for near-integrated time series, *Econometrica* 56, 1021–1043.
[61] Phillips, P. C. B. (2023) Estimation and inference with near unit roots, *Econometric Theory* 39, 221–263.
[62] Phillips, P. C. B. and B. E. Hansen (1990) Statistical inference in instrumental variables regression with I(1) processes, *Review of Economic Studies* 57, 99–125.
[63] Phillips, P. C. B. and C. S. Kim (2007) Long run covariance matrices for fractionally integrated processes, *Econometric Theory* 23, 1233–1247.
[64] Phillips, P. C. B. and T. Magdalinos (2007) Limit theory for moderate deviations from a unit root, *Journal of Econometrics* 136, 115–130.
[65] Protter, P. E. (2005) *Stochastic Integration and Differential Equations*, 2nd Edn., Springer, Berlin.
[66] Robinson, P. M. (1994) Time series with strong dependence. In *Advances in Econometrics, Econometric Society 6th World Congress. Vol. 1* (Sims, ed.), Cambridge University Press, Cambridge.
[67] Robinson, P. M. (1994) Semiparametric analysis of long memory time series, *Annals of Statistics* 22(1), 515–539.
[68] Rozanov, Yu. A. (1967) *Stationary Random Processes*, Holden Day, San Francisco.
[69] Saikkonen, P. (1991) Asymptotically efficient estimation of cointegration regressions, *Econometric Theory* 7, 1–21.

[70] Samorodnitsky, G. (2016) *Stochastic Processes and Long Range Dependence*, Springer International, Switzerland.

[71] Skorokhod, A. V. (1956), Limit theorems for stochastic processes, *Theory of Probability and its Applications* I(3), 261–290.

[72] Skorokhod, A. V. (1957), Limit theorems for stochastic processes with independent increments, *Theory of Probability and its Applications* II(2), 138–171.

[73] Sowell, F. (1990) The fractional unit root distribution, *Econometrica* 58(2), 495–505.

[74] Sowell, F. (1992) Maximum likelihood estimation of stationary univariate fractionally integrated time series models. *Journal of Econometrics* 53, 165–188.

[75] Stock, J. H. and M. W. Watson (1993) A simple estimator of cointegrating vectors in higher order integrated systems, *Econometrica* 61, 783–820.

[76] Sutcliffe, J., S. Hurst, A. G. Awadallah, E. Brown, and K. Hamed (2016) Harold Edwin Hurst: the Nile and Egypt, past and future, *Hydrological Sciences Journal* 61(9), 1557–1570.

[77] Taqqu, M. S. (1975) Weak convergence to fractional Brownian motion and to the Rosenblatt process, *Z. Wahrscheinlichskeitstheorie Verw. Geb.* 31, 287–302.

[78] Taqqu, M. S. (2003) Fractional Brownian motion and long-range dependence, Chapter 1 of [24].

[79] Tousson, O. (1925) Memoire sur l'Histoire du Nil, *Memoires de l'Institut d'Egypte*, Vol. 8, Imprimeries de l'Institut Francais, Cairo.

[80] Wang, Q., Y-X. Lin, and C. M. Gulati (2003) Asymptotics for general fractionally integrated processes with applications to unit root tests, *Econometric Theory* 19, 143–164.

[81] Wu, W. B. and Shao, X. (2006). Invariance principles for fractionally integrated nonlinear processes. *IMS Lecture Notes–Monograph Series: Recent Developments in Nonparametric Inference and Probability* 50, 20–30.

[82] Yaglom, A. M. (1962) *Introduction to the Theory of Stationary Random Functions*, Prentice-Hall, Englewood Cliffs, NJ.

Index

The manufacturer's authorised representative in the EU for product safety is Oxford University Press España S.A. of el Parque Empresarial San Fernando de Henares, Avenida de Castilla, 2 – 28830 Madrid (www.oup.es/en or product.safety@oup.com). OUP España S.A. also acts as importer into Spain of products made by the manufacturer.

www.ingramcontent.com/pod-product-compliance
Lightning Source LLC
Chambersburg PA
CBHW051401080825
30821CB00005B/35/J